20/5/08

GC/MS

GC/MS

A Practical User's Guide

Second Edition

MARVIN C. McMASTER

WILEY-INTERSCIENCE

A JOHN WILEY & SONS, INC., PUBLICATION

Library of Congress Cataloging-in-Publication Data:

McMaster, Marvin C.
    GC/MS : a practical user's guide. – 2nd. ed. / Marvin C. McMaster.
        p. cm.
    Includes index.
    ISBN 978-0-470-10163-6 (cloth/cd)
1. Gas chromatography. 2. Mass spectrometry. I. Title.
    QD79.C45M423 2007
    543′.85–dc22

                                            2007027370

Printed in the United States of America
10 9 8 7 6 5 4 3 2 1

*To the memory of
Chris McMaster
my son, my illustrator,
my partner,
and my brother in Christ*

CONTENTS

PREFACE

This book arose out of the need for a textbook for an extension course I teach at the University of Missouri-St. Louis. I had been searching for a practical guide for using and maintaining a GC/MS System to help my students drawn from university and company laboratories in our area. I have sold and supported HPLC, GC/MS, and other analytical systems for a number of years, so the course material and slides were created from my notes and experiences. I wrote the text while my son, Christopher, translated my drawings into the illustrations in this book before he pass away from the ravages of Muscular Dystrophy eight years ago.

This second addition has been updated with information on new advances in gas chromatography and mass spectrometry. This handbook is presented in sections because I believe it is easier to learn this way.

Part I presents a comparative look at gas chromatography/mass spectrometry (GC/MS) and competitive instrumentation. Then an overview of the components of a generic GC/MS system is provided. Finally, I discuss how to set up a system and perform an analysis run that provides the information you need.

After obtaining some hands-on experience, Part II on optimization provides information on tuning and calibration of the mass spectrometer, cleaning, troubleshooting problems, processing information, and interfacing to other analytical and data systems; that is, getting the whole system up and running, keeping it up, and getting useful information.

Part III provides information on the use of GC/MS in research, environmental, and toxicology laboratories, as well as more esoteric applications in space science and hazardous materials detection in the field. GC/MS has become the gold standard for definitive chemical analysis. Although quadrupole mass spectrometers predominately are used in commercial laboratories, there is a growing use of ion trap, time-of-flight, and hybrid MS/MS systems and these are discussed briefly. Magnetic sector systems, which dominated the early mass spectrometry growth, are making a resurgence along with Fourier transform GC/MS in accurate mass determination required for molecular formula and structure reporting in chemical publication, and these are discussed next.

As I taught courses I found myself moving from slide projectors to overhead projection of slides from Microsoft PowerPoint presentations. I decided to include a CD in the book with a microsoft PowerPoint slide presentation as well as tables, FAQs, etc. so a lecturer would not have to reinvent the wheel and the student could slide the CD in a computer and self-study the material. To assist in making this a self-learning tool, I went back and carefully annotated each slide.

I hope you will enjoy this book and find it as useful a reference tool for your laboratory and classroom as I have.

MARVIN C. MCMASTER

Florissant, Missouri
October 2007

PART I
A GC/MS PRIMER

1

INTRODUCTION

The combination of gas liquid chromatography (GC) for separation and mass spectrometry (MS) for detection and identification of the components of a mixture of compounds is rapidly becoming the definitive analytical tool in the research and commercial analytical laboratory. The GC/MS systems come in many varieties and sizes depending on the work they are designed to accomplish. Since the most common analyzer used in modern mass spectrometers is the quadrupole, we will focus on this means of separating ion fragments of different masses. Discussion of ion trap, time-of-flight, Fourier transform mass spectrometry (FTMS), and magnetic sector instruments will be reserved for latter sections in the book.

The quadrupole operational model is the same for bench top production units and for floor standing research instruments. The actual analyzer has changed little in the last 10–12 years except to grow smaller in size. High vacuum pumping has paralleled the changes in the analyzer, especially in the high efficiency turbo that have shrunk to the size of a large fist in some systems. Sampling and injection techniques have improved gradually over the last few years.

The most dramatic changes have been in the area of control and processing software and data storage capability. In the last 10 year, accelerating computer technology has reduced the computer hardware and software system shipped

GC/MS: A Practical User's Guide, Second Edition. By Marvin C. McMaster
Copyright © 2008 John Wiley & Sons, Inc.

with the original system to historical oddities. In the face of newer, more powerful, easier to use computer systems, the older DEC 10, RTE (a Hewlett-Packard minicomputer GC/MS control system) and Pascal-based control and data processing systems seem to many operators to be lumbering, antiquated monstrosities.

The two most common reasons given for replacing a GC/MS system is the slow processing time and the cost of operator training. This is followed by unavailability of replacement parts as manufacturers discontinue systems. The inability of software to interface with and control modern gas chromatographic and sample preparation systems is the final reason given for replacement.

Seldom, if ever, is the complaint that the older systems do not work, or that they give incorrect values. In many cases, the older systems appear better built and more stable in day-to-day operation than newer models. Many require less cleaning and maintenance. This has lead to a growing market for replacement data acquisition and processing systems. Where possible, the control system should also be updated, allowing access to modern auxiliary equipment and eliminating the necessity for coordinating dual computers of differing age and temperaments.

Replacement of older systems with the newest processing system on the market is not without its problems. Fear of loss of access to archived data stored in outdated, proprietary data formats is a common worry of laboratories doing commercial analysis.

1.1   WHY USE GC/MS?

Gas liquid chromatography is a popular, powerful, reasonably inexpensive, and easy-to-use analytical tool. Mixtures to be analyzed are injected into an inert gas stream and swept into a tube packed with a solid support coated with a resolving liquid phase. Absorptive interaction between the components in the gas stream and the coating leads to a differential separation of the components of the mixture, which are then swept in order through a detector flow cell. Gas chromatography suffers from a few weaknesses such as its requirement for volatile compounds, but its major problem is the lack of definitive proof of the nature of the detected compounds as they are separated. For most GC detectors, identification is based solely on retention time on the column. Since many compounds may possess the same retention time, we are left in doubt about the nature and purity of the compound(s) in the separated peak.

The mass spectrometer takes injected material, ionizes it in a high vacuum, propels and focuses these ions and their fragmentation products through a magnetic mass analyzer, and then collects and measures the amounts of each selected ion in a detector. A mass spectrometer is an excellent tool for clearly identifying the structure of a single compound, but is less useful when presented with a mixture.

The combination of the two components into a single GC/MS system forms an instrument capable of separating mixtures into their individual components, identifying, and then providing quantitative and qualitative information on the amounts and chemical structure of each compound. It still possesses the weaknesses of both components. It requires volatile components, and because of this requirement, has some molecular weight limits. The mass spectrometer must be tuned and calibrated before meaningful data can be obtained. The data produced has time, intensity, and spectral components and requires a computer with a large storage system for processing and identifying components. A major drawback of the system is that it is very expensive compared to other analytical systems. With continual improvement, hopefully the cost will be lowered because this system and/or the liquid chromatograph/mass spectrometry system belong on every laboratory bench top used for organic or biochemical synthesis and analysis.

Determination of the molecular structure of a compound from its molecular weight and fragmentation spectra is a job for a highly trained specialist. It is beyond the scope and intent of this book to train you in the interpretation of compound structure. Anyone interested in pursuing that goal should work through Dr. McLafferty's book listed in Appendix E, then practice, practice, practice. Chapter 12 is included to provide tools to let you evaluate compound assignments in spectral databases. It uses many of the tools employed in interpretation, but its intent is to provide a quick check on the validity of an assignment.

1.2 INTERPRETATION OF FRAGMENTATION DATA VERSUS SPECTRAL LIBRARY SEARCHING

How do we go about extracting meaningful information from a spectra and identify the compounds we have separated? A number of libraries of printed and computerized spectral databases are available to us. We can use these spectra to compare both masses of fragments and their intensities. Once a likely match is found, we can obtain and run the same compound on our instrument to confirm the identity both by GC retention time and mass spectra. This matching is complicated by the fact that the listed library spectra are run on a variety of types of mass spectrometers and under dissimilar tuning conditions. However, with modern computer database searching techniques, large numbers of spectra can be searched and compared in a very short time. This allows an untrained spectroscopist to use a GC/MS for compound identification with some confidence. Using these spectra, target mass fragments characteristic of each compound can be selected, allowing its identification among similarly eluting compounds in the chromatogram.

Once compounds have been identified, they can be used as standards to carry out quantitative analysis of mixtures of compounds. Unknown compounds found in quantitative analysis mixtures can be flagged and identified by spectral comparison using library searching. Spectra from scans at chromatography peak fronts and tails can be used to confirm purity or identify the presences of impurities.

1.3   THE GAS CHROMATOGRAPH/MASS SPECTROMETER

From the point of view of the chromatographer, the gas chromatograph/mass spectrometer is simply a gas chromatograph with a very large and very expensive detector, but one that can give a definitive identification of the separated compounds. The sample injection and the chromatographic separation are handled in exactly the same way as in any other analysis. You still get a chromatogram of the separated components at the end. It is what can be done with the chromatographic data that distinguishes the mass spectral detector from an electron capture or a flame ionization detector.

The mass spectrometrist approaches the GC/MS from a different point of view. The mass spectrum is everything. The gas chromatograph exists only to aid somewhat in improving difficult separations of compounds with similar mass fragmentations. The only true art and science to him or her is in the interpretation of spectra and identification of molecular structure and molecular weight.

The truth, of course, lies somewhere in between. A good chromatographic separation based on correct selection of injector type and throat material, column support, carrier gas and oven temperature ramping, and a properly designed interface feeding into the ion source can make or break the mass spectrometric analysis. Without a properly operating vacuum system, ion focusing system, mass analyzer, and ion detector, the best chromatographic separation in the world is just a waste of the operator's time. It is important to understand the components that make up all parts of the GC/MS system in order to keep the system up, running, and performing in a reproducible manner.

1.3.1 A Model of the GC/MS System

There are a number of different possible GC/MS configurations, but all share common types of components. There must be some way of getting the sample into the chromatogram, an *injector*. This may or may not involve sample purification or preparation components. There must be a *gas chromatograph* with its carrier gas source and control valving, its temperature control oven and microprocessor programmer, and tubing to connect the injector to the column and out to the mass spectrometer interface. There must be a *column* packed with support and coated with a stationary phase in which the separation occurs. There must be an *interface* module in which the separated compounds are transferred to the mass spectrometer's ionization source without remixing. There must be the *mass spectrometer* system, made up of the ionization source, focusing lens, mass analyzer, ion detector, and multistage pumping. Finally, there must be a *data/control* system to provide mass selection, lens and detector control, and data processing and interfacing to the GC and injector (see Fig. 1.1).

The injector may be as simple as a septum port on top of the gas chromatograph through which a sample is injected using a graduated capillary syringe. In some cases, this injection port is equipped with a trigger that can start the oven temperature ramping program and/or send a signal to the data/control system to begin acquiring data. For more complex or routine analysis, injection can be made from an autosampler allowing multiple vial injections, standards injection, needle washing, and vial barcode identification. For crude samples that need preinjection processing, there are split/splitless injectors, throat liners with different surface geometry, purge and trap systems, headspace analyzers, and cartridge purification systems. All these systems provide sample extraction, cleanup, or volatilization prior to the introduction of analytical sample onto the gas chromatographic column.

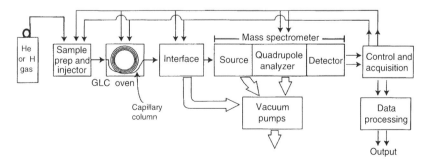

FIGURE 1.1 A typical GC/MS system diagram.

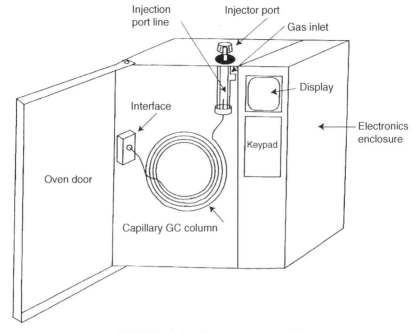

FIGURE 1.2   Gas chromatograph.

The gas chromatograph, Figure 1.2, is basically a temperature-controlled oven designed to hold and heat the GC column. Carrier gas, usually either nitrogen, helium, or hydrogen, is used to sweep the injected sample onto and down the column where the separation occurs and then out into the mass spectrometer interface.

The interface may serve only as a transfer line to carry the pressurized GC output into the evacuated ion source of the mass spectrometer. A jet separator interface can also serve as a sample concentrator by eliminating much of the carrier gas. It can permit carrier gas displacement by a second gas more compatible with the desired analysis, that is, carbon dioxide for chemically induced (CI) ionization for molecular weight analysis. It can be used to split the GC output into separate streams that can be sent to a secondary detector for simultaneous analysis by a completely different, complimentary method.

The mass spectrometer has three basic sections: an ionization chamber, the analyzer, and the ion detector (Fig. 1.3).

In the evacuated ionization chamber, the sample is bombarded with electrons or charged molecules to produce ionized sample molecules. These are swept into the high vacuum analyzer where they are focused electrically then selected in the quadrupole rods. The direct current (dc) signal charging

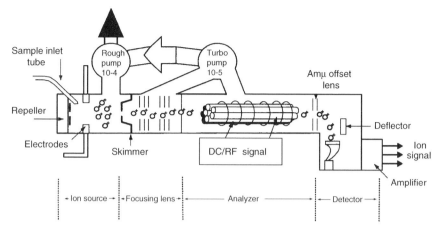

FIGURE 1.3   Quadrupole mass spectrometer.

apposing poles of the quadrupole rods creates a standing magnetic field in which the ions are aligned. Individual masses are selected from this field by sweeping it with a radio frequency (RF) signal. As different dc/RF frequencies are reached, different mass/charge ratio (m/z) ions are able to escape the analyzer and reach the ion detector. By sweeping from higher to lower frequency, the available range of m/z ions are released one at a time to the detector, producing a mass spectrum.

On entering the ion detector, the ions are deflected onto a cascade plate where the signal is multiplied and then sent to the data system as an ion current versus m/z versus time. The summed raw signal can be plotted against time as a total-ion chromatogram (TIC) or a single-ion m/z can be extracted and plotted against time as a single-ion chromatogram (SIC). At a single time point, the ion current strength for each detected ion fragment can be extracted and plotted over an m/z mass range, producing a mass spectrum. It is important always to remember that the data block produced is three dimensional: (m/z) versus signal strength versus time. In most other detectors, the output is simply signal strength versus time.

1.3.2   A Column Separation Model

Separation of individual compounds in the injected sample occurs in the chromatographic column. The typical gas chromatographic column used for GC/MS is a long, coiled capillary tube of silica with an internal coating of a either a viscous liquid such as carbowax or a wall-bonded organic phase.

The injected sample in the carrier gas interacts with this stationary organic phase and equilibrium is established between the concentration of each

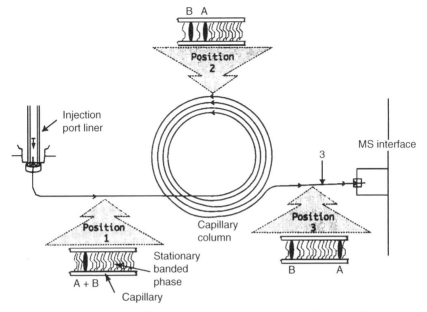

FIGURE 1.4   Chromatographic column separation model.

component in the gaseous and solid phases. As fresh carrier gas flushes down the column, each compound comes off the stationary phase at its own rate. Separations increase after many interactions down the length of the column; then each volatile component comes off the column end and into the interface (Fig. 1.4).

Both the injector and the column can be heated to aid in compound removal since not all components of the injected sample are volatile at room temperature. The column oven allows programmed gradient heating of the column. Temperatures above 400°C are avoided to prevent thermal degradation of the sample.

Moving down the column, the injection mixture interacts with the packing. Separation is countered by remixing due to diffusion and wall interactions. Finally, each compound emerges into the interface as a concentration disc, tenuous at first, then rising to a concentration maxima and then dropping rapidly as the last molecules comes off. If we were to run this effluent into an ultraviolet (UV) detector, we would see a rapidly rising peak reach its maximum height then fall again with a slight tail.

Ideally, each compound emerges as a disc separated from all other discs. In actual separations of real samples, perfect separation is rarely achieved. Compounds of similar chemical structure and physical solubility are only poorly resolved and coelute. In a chromatographic detector, they appear as overlapping or unresolved peaks. Something else must be done to prove their presence, to identify their structure, and to quantitate the amounts of each compound.

1.3.3   GC/MS Data Models

The simplest data output from the mass spectrometer analyzer is a measurement of total-ion current strength versus time, a TIC (Fig. 1.5).

   This is basically a chromatographic output representing a summation of the signal strength of all the ions produced by the mass spectrometer at a given time. The chromatogram produced is similar in appearance to a UV chromatogram with peaks representing the chromatographic retention of each component present. In a UV detector, however, you would see only the

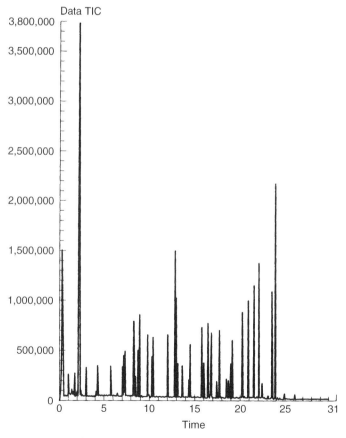

FIGURE 1.5   Total-ion chromatogram.

compounds that absorb UV light at the selected wavelength. In the mass spectrometer, any compound capable of being ionized and forming fragments would be detected. The mass spectrometer serves as a universal chromatographic detector.

The actual data output content is much more complex. If the mass spectrometer is in the scanning (SCAN) mode, the analyzer voltage is being changed continuously and repeatedly over a selected mass range. Different ion masses are reaching and being detected by the detector. Information is coming out each moment on the exact position of the analyzer. After calibration and combination with the ion concentration information, this provides the molecular mass and amounts of each ion formed. After these data are computer massaged, we receive a three-dimensional block of data whose coordinates are elapsed time, molecular mass (m/z), and ion concentration (Fig. 1.6).

Viewing this block of data on a two-dimensional display such as a integrator or a CRT screen while trying to extract meaningful information is nearly impossible. A three-dimensional projection on a screen can be made but is not particularly useful for extracting information. It does provide an overall topologically view of the data, which is useful for finding trends in the data set.

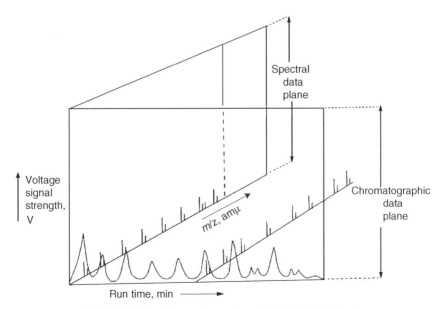

FIGURE 1.6   Three-dimensional GC/MS data block.

If we select a data cut at a single molecular mass, we can produce a SIC similar to that produced by a UV detector tracing at a single wavelength (Fig. 1.7).

The series of peaks produced represent the concentration of ions of the selected molecular mass present throughout the chromatographic run. Compounds that do not form an ion with this mass will not be present in the single-ion chromatogram. Comparison with the TIC shows a much simplified chromatogram, but all peaks in the SIC are present in the TIC.

An SIC can also be produced by running the GC/MS in a fixed-mass mode in which the analyzer is parked at a given molecular mass position through out the chromatographic run. This single-ion monitoring (SIM) mode has an additional advantage. Because the analyzer is continuously analyzing for only a single ion, the summed ion yield is much higher and detection limits for this

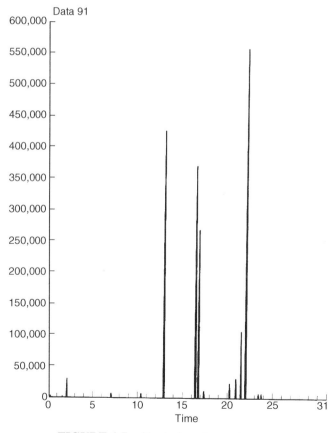

FIGURE 1.7 Single-ion chromatogram.

ion are much lower. The mass spectrometer becomes a much more sensitive detector, but only for compounds producing this mass fragment. Other compounds lacking this fragment ion will be missed. A good detector for trees instead of forests—for trace analysis of minor contaminants.

Going back to our original three-dimensional block of data in Figure 1.6, we can select a data cut at a given time point which will provide us with a display of molecular mass versus ion concentration called a mass fragment spectra or simply a mass spectra (Fig. 1.8).

Generally, these data are not displayed as an ion continuum. The ion mass around a unitary mass is summed within a window and displayed as a bar graph with 1-amu increments on the *m/z* mass axis, as shown in Figure 1.8.

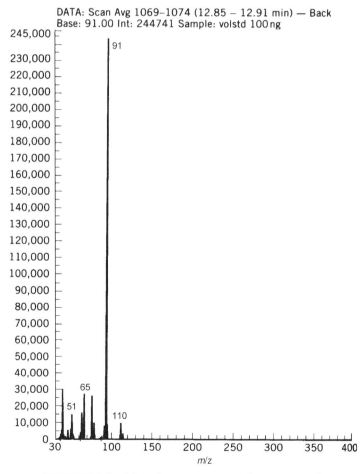

FIGURE 1.8   Mass fragment spectra (mass spectra).

The mass spectrum of a resolved compound is a record of the fragmentation pattern of this compound under a given set of experimental conditions. It is characteristic of that compound and can be used to definitively identify the chemical nature of that compound. In the same or a similar instrument under the same tuning conditions, this compound will always give the same fragments in the same ion concentration ratios. Libraries of compound fragmentation patterns can be created and searched to identify compounds by comparison with known fragmentations. Further decomposition of isolated fragments can be studied in triple quadrupole GC/MS/MS systems to identify fragmentation pathways useful in determining structures of unknown compounds.

There is a lot of arm waving involved with the statement "under a given set of experimental conditions." Different ionization methods and voltages will affect the fragmentation ions produced. Under certain conditions, only a single major ion is produced, the molecular ion. It is formed by the original molecule losing an electron to form this ion radical, whose mass is equal to the molecular weight of the compound, a very useful number to have in identifying compounds.

Changes in the geometry, calibration, cleanliness, and ion detector age of the mass spectrometer can all produce variations in the fragmentation pattern and especially in the ion concentration ratios. Variations in the chromatographic conditions can lead to overlapping peaks and change the relative heights in the fragmentation pattern. Learning and controlling these is what converts GC/MS from a science to an art. All of this has lead to a proliferation of instrument types and calibration standards attempting to tame these variables.

1.4 SYSTEMS AND COSTS

Instrument system costs are not widely advertised by manufacturers unless you work for the federal government and are buying off a Government Service Administration price list. To come up with even ballpark figures, I have talked to former customers who have recently purchased systems and have talked to manufacturers at technical meeting. The numbers in Table 1.1 represent an educated guess at 2005 system pricing.

In the past, systems could be divided in two basic types, floor standing research systems designed for the mass spectrometry research laboratory and desktop systems designed for both commercial analytical laboratories and the university analytical chemistry laboratory. A new product niche has opened in the last 10 years. These systems are simpler, easier to maintain and calibrate,

TABLE 1.1   Estimated GC/MS System Prices

MS type	GC system	MS only	GC/MS	AS/GC/MS data
Quadrupole	Production	N/A	N/A	$36,800[a]
	Benchtop	$72,000	$8000	$90,000
	Research	$90,000	$96,000	$111,000
	Triple quadrupoles	N/A	N/A	$250,000–$500,000
	Used quadrupole	>$3000	>$5000	$3000–$50,000
Ion trap	Ion detector	$65,000	$72,000	$82,000
	Research MS/MS	N/A	N/A	$450,000

[a]No autosampler.

and aimed at the quality control and analytical testing laboratories. They are advertised at a third of the price of desktop system of 12 years ago. The jury is still out on these, but some of their manufacturers have good pedigrees and track records.

I have included pricing on GC/MS/MS systems and on research and desktop ion trap GC/MS systems for comparison with the quadrupoles because many users consider these the analytical systems of the future. The three-dimensional and linear ion traps seem to be simpler, more sensitive, ideal systems for MS/MS studies. If the future is truly toward smaller, more compact systems, the linear ion trap GC/MS system may lead the way because of its versatility and increased sensitivity for trace component studies.

Overall, there definitely is a trend toward lower pricing and ease of operation. This will make systems more available to the average research investigator and commercial laboratory.

There is a growing market for older GC/MS systems because of price and the availability of upgraded data systems, both from GC/MS manufacturer and from third-party sources. It is true that the old data system is usually the worst part of the older system; computer technological advances having left them in the dust. They are difficult to learn, hard to use, and very difficult to connect into modern data networks, since their data formats are obsolete or on the verge of becoming obsolete.

Pumping and analyzer section almost always work. Ion detectors and data systems can generally be replaced if necessary. Once retrofitted, these systems usually perform like champs.

However, be aware that there are some real old dogs out there. Systems that were never very good and no amount of retrofitting will improve them. Systems without butterfly valves in the oil pump that dump pump oil into the analyzer in case of power failures, systems whose manufacturers have disappeared into the night, or-one-of-a-kind systems in which no two systems have the same control inputs or detector outputs. I know because I have demonstrated replacement data systems on all of these. Let the buyer beware!

When retrofits work, they are often great buys. I had a customer who purchased a hardly used GC/MS from a hospital for $25,000, added a modern data/control system for $22,500, and had a state-of-the-art system for under $50,000. I know a production facility just getting started that bought 12-year-old systems for $3000 each, modernized the data system, networked them, and ran them day and night until they could afford to replace them with 20 newer systems. They purchased bare systems, without a processing and control computer, and moved the existing data/control systems to each new instrument as they purchased them. Operator retraining was negligible as well as system switchover time.

The key to buying older system is to buy one made by a company that was successful when the system was sold and is still successful. Talk to someone who has used or is still using the same type of instrument. Find out what he thinks about it—its strengths and limitations.

1.5   COMPETITIVE ANALYTICAL SYSTEMS

What other analytical systems do you need to consider when selecting an instrument to use in your research? Table 1.2 gives us an idea of the size of the analytical systems market in 2006.

If you need the definitive identification provided by a GC/MS system, there are few competitive system and none at the same relatively mature state of development. On the horizon are a few contenders for the crown. LC/MS has a fairly broad application potential, others fit better in specific analytical niches.

1.5.1   Liquid Chromatography/Mass Spectrometry (LC/MS)

The high performance liquid chromatograph (HPLC) connected to the mass spectrometer in my opinion offers the best potential as a general MS

TABLE 1.2   2006 North American Chromatography System Sales

GC/MS	226 million	4.9% growth
LC/MS	154 million	5.0% growth
HPLC	1206 million	5.8% growth
GC	326 million	2.0% growth
SPE	112 million	6.0% growth
IC	106 million	5.7% growth

LC/GC Magazine 2007 Media Planner online.

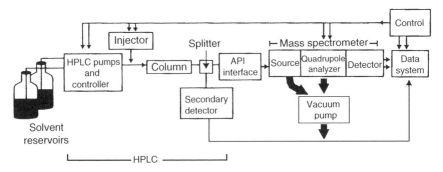

FIGURE 1.9 LC/MS system diagram.

instrument for the laboratory. These LC/MS systems aimed at the production as well as the general research laboratory began to appear on the market about 10 years ago. They now claim major improvements in ease of maintenance and operator training, in calibration stability, in interface flexibility, and in system pricing (Fig. 1.9).

Chromatographically, the HPLC offers flexibility in media and in isocratic and solvent gradient separation technology. Almost anything that can be dissolved can be separated, generally without much sample preparation or derivatization. Large molecules such as proteins and restriction fragments can be separated and analyzed using electrospray techniques.

Limitations to using the technique are due to the current failure of LC/MS systems to provide molecular ion fragmentation without going to a LC/MS/MS system and greatly increasing the cost and experience need to use these systems. The chemical ionization used in atmospheric pressure chemical ionization (APCI) interfaces is a soft ionization and does not usually fragment the molecular ion. LC/MS currently can be used to provide retention times and molecular weights of the separated materials but not definitive compound identification. Existing analytical techniques and calibration standards are just appearing; few have been accepted by and approved by regulatory agencies. Price and reliability are still considerations for general laboratory application. Existing spectral libraries may require modification to be used with LC/MS/MS analysis and definitely will need additional compounds added to them.

1.5.2 Capillary Zone Electrophoresis/Mass Spectrometry (CZE/MS)

Another research tool of growing popularity, capillary electrophoresis interfaced with a mass spectrometer offers a powerful but limited tool for analytical separations (Fig. 1.10).

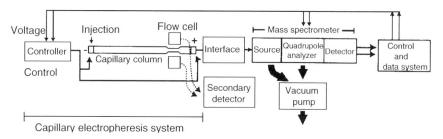

FIGURE 1.10   CZE/MS system diagram.

Capillary electrophoresis uses electromotive force to separate charged molecules in a capillary column filled with buffer or buffer containing gel. A very strong electrical voltage potential is applied to either end of the column. Ionized compounds moved toward the electrode with the opposite charge at a rate dependent on their size and charge strength. It is designed to work with very small amounts of material and delivers a very concentrated compound disc to the mass spectrometer interface. Very high efficiency separation can be achieved. It has proved very useful for analyzing multiple charged molecules such as proteins and DNA restriction fragments when combined with an electrospray MS interface. Limitations for general application have been injector design problems, necessity to work with very high voltages and high concentrations of volatile buffer, and problems eluting samples into the MS interface. Current systems cost, high levels of maintenance, and calibration stability problems have prevented this technique from wider application, but these appear to be coming under better control. Like LC/MS, there are few approved methods for production applications.

1.5.3   Supercritical Fluid Chromatography/Mass Spectrometry (SFC/MS)

Widely considered to be only a laboratory curiosity, this technique has been adopted by manufacturers of flavors, essential oils, and may be developed into a useful environmental analysis tool. One of its most attractive features is its use in combination with supercritical fluid sample extraction in automated sample preparation and analysis.

In SFC/MS, gases such as carbon dioxide can be used in their supercritical fluid state as a mobile phase for separation of injected material on a normal phase HPLC column. Equipment from the injector to the detector interface must be operated under the pressures needed to maintain the gas in its supercritical state (Fig. 1.11).

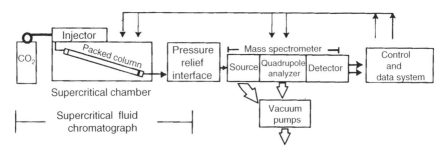

FIGURE 1.11   SFC/MS system diagram.

The major advantage to the techniques is that mobile phase is dispersed simply by reducing pressure in the ionizing interface. Except for minor carbon dioxide contamination, there is almost no solvent background in the mass spectrometer operation. Limitations are the requirements for high pressure chromatographic operation, limited mobile phase selection, and lack of availability of commercial equipment and methodology. The latter two problems may quickly disappear if a specific application develops that only SFC/MS can answer.

2

SAMPLE PREPARATION AND INTRODUCTION

The mass spectrometer is designed to analyze only very clean materials. Even solvent can interfere with fragmentation pattern identification. We attach a gas chromatography to separate materials and, obviously, it can clean samples. But, separation and sample cleanup on real-world sample exceeds the capacity of the device.

Before introducing the sample into the gas chromatograph, some form of sample preparation is needed. If we can combine the sample preparation and introduction into a single, automated apparatus, we have achieved our purpose. In this chapter, we will look at methods for direct injection into the mass spectrometer, gas chromatography injection techniques, and extraction methods for freeing our compounds of interest from various environmental matrices.

Sample preparation for GC/MS injection falls under three main categories: volatile organics, extractable, and special sample preparation. Volatiles are usually uncharged organics that are injected directly into the GC column. These may be collected and concentrated in a headspace analyzer before injection. Purge-and-trap apparatus is used to remove organics by gas sparging out of solution, trap them in a retention column, and then inject these compounds onto the GC column by reversing the gas flow and heating the trapping column. Volatile organics may be dissolved and injected in a carrier

GC/MS: A Practical User's Guide, Second Edition. By Marvin C. McMaster
Copyright © 2008 John Wiley & Sons, Inc.

solvent or extracted from an aqueous mixture with a carrier solvent, dried to remove moisture, and injected. Extractable are charged compounds that have to be neutralized before extraction and are usually dried before injection. Special preparation samples must be treated before dissolving them for injection. Often these are highly charged compounds that cannot be neutralized and extracted. They may require preparation of volatile derivatives or other treatment to aid in detection.

2.1 DIRECT SAMPLE INJECTION INTO THE MASS SPECTROMETER

Many GC/Ms systems have a port for a direct insertion probe (DIP) on which a sample may be inserted directly into the mass spectrometer source for ionization and analysis. The sample can be introduced as a drop of liquid, a solid, as a film dissolved in a volatile solvent, or as an emulsion or suspension in an activating compound for fast atom bombardment (FAB).

2.1.1 Direct Insertion Probes and Fast Atom Bombardment

The DIP port is a vacuum lock through which the probe can be inserted without disturbing the analyzer vacuum. Pendant off this DIP port is a connection line for attaching a bulb containing calibration gas. The probe itself is a metal tube, usually equipped with an electrical heating unit, which ends in a slanting sample face that is inserted into the ionizing source beam. The sample is volatilized by the vacuum system or by programmed heating of the probe heating element (Fig. 2.1).

The FAB technique is used for ionizing nonvolatile solids. The sample is ground into an emulsion in a viscous liquid such as glycerin. A drop of this emulsion is placed on the probe face and inserted into the source through the port. The sample is bombarded with heavy ions, often Cesium, from a special ionization source. The droplet absorbs the energy, explodes, and throws some of the sample into the source cavity where it is ionized and swept into the analyzer. If this sounds like a messy procedure, it is. Fast atom bombardment sources require frequent cleaning. But, FAB's usefulness for nonvolatile samples is great enough to make it a very popular technique.

The literature has reported attempts at combining GC separation with FAB in a technique called flow FAB. The GC effluent is eluted into a separator interface to remove part of the carrier gas and to introduce a FAB solvent. The solution or suspension is then sprayed into the evacuated FAB ionization

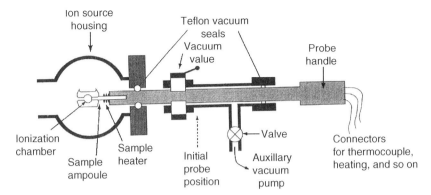

FIGURE 2.1   Direct insertion probe.

source. The technique has also been used in time-of-flight mass spectrometers to introduce absorbing dyes that are excited with laser sources to aid in sample ionization.

2.1.2   Headspace Analyzers

Another direct injection technique used in GC/MS laboratories is a headspace analyzer. Sample is introduced into an evacuated chamber, which is then sealed. The sample can then be vented directly into the evacuated mass spectrometer source where each component can be analyzed using SIM mode. A distinctive mass ion is chosen for each component of interest, and these ions are monitored sequentially using step scan analysis. I have seen this technique being used for monitoring automobile exhaust gas ratios over days, weeks, and months.

Alternatively, a headspace analyzer can be set up to feed a gas chromatograph. The sample is introduced into an evacuated chamber through a vacuum port and the volatilized sample components are swept into the GC with carrier gas where they concentrate on the column head before separation and elution.

2.2   SAMPLE PURIFICATION

Nature has a habit of creating complex mixtures. In analyzing these, we turn to techniques that are compatible with the constraints of the mass spectrometer. Injection solvent and carrier gas removal and sample concentration are two of the major problems that need to be solve before the mass spectrometer could be connected to a gas chromatographic system effluents. They have to be solved also when purifying samples for GC/MS analysis.

The problem with purification using extractions is the difficulty of getting rid of the extracting solvent and the carrier gas before introducing the sample into the mass spectrometer. The mass spectrometer is an excellent analyzer for trace amounts of unremoved solvents. For instance, is you extract a soil sample suspended in acidified water with methylene chloride and then inject it into a nonpolar GC column, you will be dealing with a severe methylene chloride peak contamination problems if you do multiple injections to increase sample load. If you put the extract in a nitrogen tube dryer, evaporate the methylene chloride, and take the sample up in methanol (assuming it will redissolve in methanol) before injecting, you will still have a methylene chloride contamination problem. But, now you have to split off the extraction methanol before it gets to the mass spectrometer source.

Both of these techniques are used, but it is best to avoid solvents additions where ever possible. A step in the right direction is techniques using preinjection cartridge columns for both extraction and concentration to minimize extraction volumes. Organics in aqueous solution can be partitioned on to these columns with the stationary phase acting in place of an organic solvent. The retained material can be eluted with a small amount of methanol or acetonitrile for injection into the gas chromatograph.

Use of polar-phase cartridges in a supercritical fluid apparatus would allow extraction with supercritical carbon dioxide. With pressure reduction directly in a carrier gas stream, most compounds could be swept directly onto the GC column. "Most" is an important qualifier since many compounds would precipitate in the cartridge on pressure reduction and only be volatilized by heating the injector port, if at all. This technique is being evaluated for automated extraction of environmental sample for direct injection into GC/MS systems.

The next purification method is to purge volatiles from aqueous samples with carrier gas directly into the GC injector port. This introduces a lot of water into the injector, but suggests the next improvement, the purge-and-trap apparatus (Fig. 2.2).

This apparatus is made of a purge tube, which may be heated with a sleeve, containing your aqueous sample and a gas purge tube. The volatile are swept by purge gas into a packed dryer column in which they are trapped until the end of the purge period. Unretained purge gas is vented from the system. For elution into the GC injector, carrier gas is swept in the reverse direction through the trap that is heated to elute the dry trapped sample directly onto the GC column.

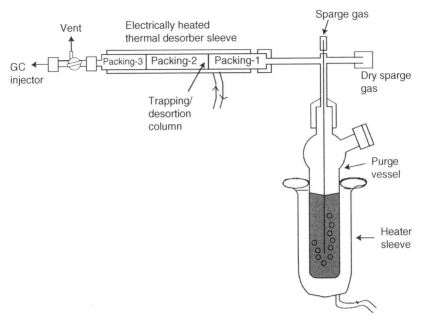

FIGURE 2.2   Purge-and-trap injection system.

2.3 MANUAL GC INJECTION

Once we a have a sample in a syringe ready for injection, we need a way of introducing it onto the gas chromatography column. The split/splitless injector has become the model for manual GC injection (Fig. 2.3).

This injector is capped with a self-sealing, replaceable septum for syringe injection, a carrier gas inlet to sweep sample into the injector body, an automated valve for sample diversion, a heated throat with a removable throat liner, and a seal fitting it to the top of the capillary column. Septum-less syringe ports have been introduced which use a spring-sealed Teflon surface to seal around the injection needle. It is important to use a specially designed blunt syringe needle with these ports. A pointed needle will score the Teflon surface and cause leaking.

Sample is vaporized in the injector throat. The split valve is used to control the amount of the sample that is allowed to enter the column. This is used primarily to prevent overloading the column. Since sample discrimination can occur during volatilization and splitting, a variety of throat liners are available that provide variations in surface area and composition to control these evaporative changes. The simplest throat liner is a plug of glass wool, but a

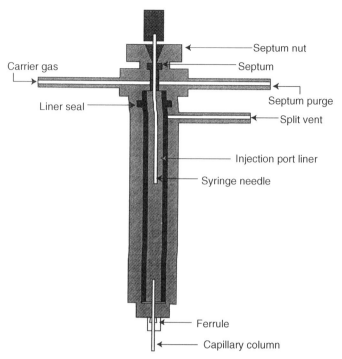

FIGURE 2.3   Split/splitless GC injector.

variety of borosilicate glass and silica-restricted tubes with constrictions are available. Specific throat liners and split times are often specified in method protocols. Liners must be cleaned daily to remove nonvolatile components from the injection.

Many systems are being equipped with injectors that allow direct liquid sample injection on to the column head. This is done primarily to prevent thermal decomposition of the sample in the heated injector throat. It also avoids sample discriminations and pressure variations associated with volatilization and split valving.

The obvious problem with on-column injections is that you are introducing nonvolatile materials onto the head of the column and they will accumulate over a period of time just as it does in an injector's throat liner. There is no good way to remove them from the column; even long period bake outs will not remove many of these compounds. They can be ignored, but many will result in long-term bleeds that can interfere with mass spectrometer operation. The simplest solution to the problem is to cut a few centimeters off the column inlet periodically, since it will have little effect on the performance of a 25–150-m capillary column.

Carrier gas pressure is the primary driving force in controlling eluting time from the capillary column. Variation in the injector pressure will lead to variations in chromatographic retention times. Automated electronic pressure control (EPC) was introduced to control this variable.

It also offers potential as gradient control variable similar to solvent programming in HPLC. By reducing the pressure at specific points in the chromatographic run, compressed areas of the separation can be allowed further interaction with the column to improve resolution. Widely separated peaks can be eluted more rapidly by increasing the pressure. Method developments using this technique have appeared in the literature, leading to a dramatic decrease in the total chromatographic run time.

2.4   AUTOMATED GC/MS INJECTION

The simplest form of injection automation is the injection trigger. This usually consists of a lever or a ring switch around the septum that is depressed when the sample is injected. It acts as a contact closure to send a signal to the oven to start to run a temperature program, to the EPC valve to run a pressure program, or to the data system to start the mass spectrometer scan and to begin acquiring data.

If you are analyzing multiple samples per day, running night and day, you will need to work three shifts or automate your sample injection. Autosamplers are robotic arms that pull a sample from a specific sample vial in a carousel and inject it into the injector body connected to the capillary column (Fig. 2.4).

Each autosampler system provides some way of washing the injection needle between injections. They may also have provisions in their programming to allow them to make repeated injections from the same vial or to make periodic returns to a series of standard vials to make calibration check injections. They may provide for positive vial identification by reading a bar code label on the vial to be sent to the data system. This number allows confirmation of the identity of the sample that has just been analyzed by the mass spectrometer. Some autosamplers provide sample carousel cooling to prevent sample degradation during standing in solution in a long series of runs.

Sample vials are usually sealed with a self-sealing septum that is penetrated by the injection needle after the vial is positioned by the arm. Sample is either placed in the needle by suction from the needle line or pushed in by hydraulic pressure on the sample surface. Most autosampler needles are filled by suction: either by a syringe connected to the needle line or by a mild vacuum on a automatic valved line.

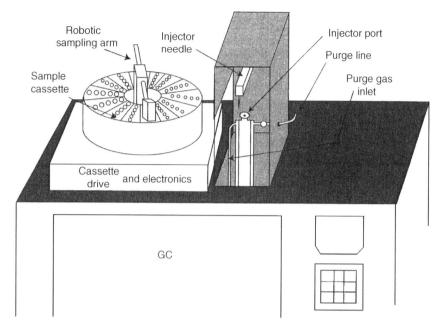

FIGURE 2.4   Autosampler diagram.

Once filled, the needle is swung into position over the injector septum. Injection is done by mechanically inserting the needle through the septum and then switching the syringe drive or valving over to a low-pressure line. Once injection is completed, the needle is removed from the septum, rinsed, possibly air dried, and positioned for the next sample.

The purge-and-trap apparatus can also be automated with an autosampler feeding the purge tube. A number of automated systems are on the market that can be loaded, programmed, and left to run unattended. Sample heating, purge gas, carrier gas, and trap exhaust valves are all automated from a programmable microprocessor base controller. Trap heating temperature and heating time are also under programmer's control.

A purge-and-trap system on the market will automate up to 16 purge tubes for liquid or solid samples. Another system uses only a single tube for liquid sample purge, but feeds samples into the purge tube from an autosampler. Obviously, some provision must be made for purge tube washout between samples to avoid cross contamination of sample. Both forms of automated purge-and-trap are specified for use in Environmental Protection Agency (EPA) approved methods.

3

THE GAS CHROMATOGRAPH

A gas chromatograph is a programmable oven designed to run GC columns. It has a microprocessor-based controller whose major purpose is to provide temperature gradient programming. A secondary function of the controller is to provide automated actuation of switching valves.

3.1 THE GC OVEN AND TEMPERATURE CONTROL

Oven temperature controllers usually have at least five programmable linear ramping segments. In creating a temperature profile program, linear ramps with differing slope rates are linked to achieve separation at different points in the chromatography run. Carrier gas flow rate and auxiliary valve switching can be set and changed at time points along the temperature ramps. The oven program is usually started with an injection signal, but can also be started from the front panel or from the mass spectrometer's computer control panel. The ideal is to arrange the system so that one run signal starts all of the run components.

Internally, the GC is made of three compartments (Fig. 1.2). The electronic enclosure provides space for the microprocessor boards with the display and keyboard on its exterior face. An unheated area holds the injector port head, injector trigger, purge gas lines, valves, and cabling. The large, insulated cube

GC/MS: A Practical User's Guide, Second Edition. By Marvin C. McMaster
Copyright © 2008 John Wiley & Sons, Inc.

of the oven, with a door making up its front face, holds the injector body and the column, and provides an exit port for the mass spectrometer interface connected to the end of the column. Secondary detectors are connected via a T-splitter line to the outlet of the capillary column with the detector body and flow cell either on top of the GC or in the unheated cable/injector area.

To shorten chromatographic run times, automated external cooling of the GC oven for temperature re-equilibration may be provided. This may be as simple as a mechanism to open the GC's door until the temperature drops. Or there maybe provisions for adiabatic cooling using compressed carbon dioxide gas, a technique called cryoblasting.

3.2   SELECTING GC COLUMNS

The typical gas chromatographic column used for GC/MS is a 25–150-m coiled capillary tube with an internal diameter of 0.25–0.75 mm. Drawn from either glass or silica, it has an activated surface and an internal coating of a viscous liquid such as Carbowax that acts as the stationary phase of the gas–liquid separation. Figure 3.1 illustrates the technique used in drawing and placing a protective epoxy coating on a typical capillary column.

Since coated columns have a slow bleed of the stationary phase, newer columns have been created with the stationary phase cross-linked and chemically bonded to the silica wall. These columns are more stable, give more reproducible results, less mass spectrometer contamination, and have longer working lives for GC/MS separations.

A wide variety of stationary phase with different chemical composition are available. The most common film or bonded phase is nonpolar material such as methyl silicone. This packing type is stable, has a high capacity, and provides a separation that parallels the compounds' boiling points. Low boilers come off first, high boilers last. This column is also described as a carbon number column since the more carbons in the compound's structure, the later it comes off the column. Other columns, such as Carbowax, phenylcyanopropyl, and trifluoropropyl, either are more polar, showing an affinity for hydrogen-bonded compounds, for compounds with functional groups, or for compounds with a high dipole moment. These columns can be used to optimize the separation of compounds that are not resolved using a simple boiling point column. Up-to-date catalogs from column suppliers usually supply guidelines for selecting a column for a specific separation.

For analysis under a method specified by a government regulatory agency, there may be little choice about the nature of the column selected for the work.

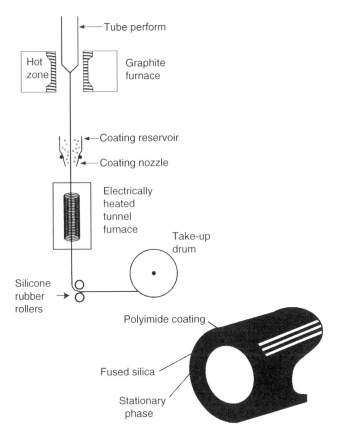

FIGURE 3.1 Preparation of the GC column.

Both coated and bonded columns are used in commercial analysis laboratory, but bonded columns are definitely the choice for new method development.

The final step in column preparation is to cover the outside of the column with a polyimide or metallic coating to provide protection against shattering (Fig. 3.1). Column coating, bonding, and packing is an art best left to the expert.

There are always some variations between columns when running standards. If reproducibility is important to your work, select a column from a reliable manufacturer and stay with it. Check the quality control with your own column standards when the column comes in, but do not jump from manufacturer to manufacturer because of price. Your time and the results you put out are too important to risk them to save a buck on a column.

If you are doing preparative GC, you may find a need for open tube, coated column. The only time to select anything but a bonded-phase, capillary column for mass spectrometry is if the method you are trying to duplicate specifies one. Try to find another method that does the same thing on a capillary or on a bonded-phase column. The efficiency and stability you gain will always be worth the effort and time.

3.3   SEPARATION PARAMETERS AND RESOLUTION

Optimum chromatographic separation is achieved when you have baseline resolution between all adjacent peaks in a reasonable run time. In order to achieve this type of separation, it is necessary to understand the variables that can be used to effect a separation.

In Figure 3.2, we have a two-compound chromatographic run made at high strip chart speed so we can measure system parameters, that can be used to define the separation.

Two parameters are measured, the retention time t_r and the peak width w of each peak. We also need to measure or know the void volume time of the column, $t_{r,0}$, the retention time of an non-retained solvent.

From these data, we can calculate a retention factor k (also called the capacity factor), a separation factor α and an efficiency factor n. Finally, using these values, we can produce a resolution equation combining all of these factors. The actual value of these numbers may be important to the theoretician, but I find them useful mainly for predicting the changes I can

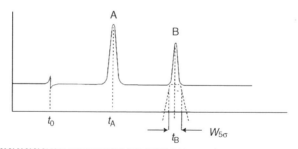

Retention factor
$$K_A = \frac{t_A - t_0}{t_0}$$

Separation factor
$$\alpha = \frac{t_B - t_0}{t_A - t_0} = \frac{k_B}{k_A}$$

Efficiency factor
$$n = 5.4 \left(\frac{t_B}{w_{\frac{1}{2}}h}\right)^2 \quad \text{or} \quad 16 \left(\frac{t_B}{W_{5\sigma}}\right)^2$$

Heigth equivalent to a theoretical plane
$$h = \frac{L}{n}$$

Resolution equation :
$$Rs = \frac{1}{4}\left(\frac{\alpha-1}{\alpha}\right)\left(\sqrt{n}\right)\left(\frac{k}{k+1}\right)$$

FIGURE 3.2   Gas chromatograph column parameters.

make in a separation during methods development or as diagnostic aids in following separation changes due to column aging over time.

Looking at the chromatogram in Figure 3.2, we can see that if $t_{r,0}$ is set to 1, $t_{r,A} \sim 0.5$, $t_{r,B} \sim 1$, and ~ 2. Since we have baseline resolution between the two peaks, we have a useable separation. However, we may not have an optimum separation if the run time is too long. It should be noted here that all peaks do not have to have baseline resolution in a useable separation. Someone has to make a decision on how perfect a separation must be to be used for a particular analysis. Reproducibility is often more important than separation perfection in the real world.

The efficiency factor n tells how sharp the peaks are and how much overlap is occurring between adjacent peaks. The sharper the peaks, the closer they can be run together, the faster they will separate, and the shorter will be the overall run time.

To measure efficiency, we must measure peak width at a given retention time. The longer a peak is on the column, the wider will be its peak width due to diffusion-induced banded spreading. The most common width measured in GC is the width at half peak height because GC peaks are very symmetrical. If your peaks tail on the back side, as they do in HPLC, you are better off using the second efficiency calculation, which measures the peak width, called the five-sigma width, at approximately 1/10 the peak height, which is much more sensitive to column changes and contamination than the halfpeak measurement. It is measured by extending each side of the peak slope until it meets the baseline, then the baseline segment formed is measured. The two methods of efficiency calculation are identical for symmetrical peaks.

Efficiency is reported in theoretical plates/meter. The larger the plate count, the higher the efficiency, and the sharper should be the separation peaks. A related value is the height equivalent to a theoretical plate, h, or the column length divided by the efficiency. In h, higher efficiency leads to smaller numbers. Plots of h versus flow velocity are used to select optimum flow rate for columns made from different packing particle diameters.

Finally, efficiency, retention, and separation factors are combined into the resolution equation, Rs. The resolution equation in Figure 3.2 shows that the retention factor term of the equation is a convergent term. It has a big initial effect on resolution, but it falls off as the factor gets larger. The efficiency factor term is a square root function. Efficient changes do not produce a linear response in resolution. Changes in the separation factor term of the resolution equation have a nearly linear effect on resolution. So what exactly are the variables that effect retention, separation, and efficiency? These will be discussed next.

3.4 GC CONTROL VARIABLES

Oven temperature is the major operational control variable followed by carrier gas pressure, which controls gas velocity. It would be nice if each variable changed only one resolution factor at a time. Some do, but many exhibit complex effects on more than one factor. Only a few variables can be used as a control variable once the column and carrier gas has been selected. The first five variables below are selected before starting the run.

Variables effecting separation are as follows:

1. *Stationary phase chemistry.* Column chemistry changes produce α effects that lead to switching of the relative peaks positions from one column type to another. On one column, the peaks may elute a, b, c; on another, they may elute a, c, b; or the two peaks may coelute. Traditional GC column packings such as Carbowax separate compounds primarily by carbon number; the more carbons in the molecule, the longer it is retained. Newer bonded-phase column with altered surface chemistry offer great potential for taking advantages of α changes to separate unresolved compounds. A number of new column types have appeared in the last few years, but have only slowly been adopted for separations. If you cannot make a separation try, change the type of column you are using. For instance, supports containing aromatic compounds should have an affinity for double bonds and aromatic compounds.

   In the past, changing columns in a GC/MS system has often been prohibited by the time needed to change a column. The high vacuum on the analyzer would first have to be broken, the column changed, and then vacuum re-established, which is time consuming. A new device introduced by Agilent allows electronic switching of the interface to mass spectrometer flow path, leaving the mass spectrometer under vacuum while the column is being changed. This promises to make column switching a usable methods development tool.

2. *Stationary phase thickness.* Thicker coats increase retention times, k, because the sample has more opportunity to interact with the stationary phase. Thick-phase columns are used for analysis of light components, thin phases for heavier components. With heavily cross-linked supports, there is inhibition of transfer through the support thickness which can be overcome with an increase in gas pressure and temperature. Be aware that nonbonded, thick-coat columns are susceptible to dramatic column failure on heating or shock as the column coat separates from the support bed and beads up.

3. *Column internal diameter.* Decreasing the column diameter increases both efficiency n and retention k. Less material channels down the column center and the ratio of gas and liquid phase favors better interaction with the column. Retention times increase as well as total run times.

4. *Column length.* The length of column has an effect on efficiency n, but the resolution equation tells us that the change is related to the square root of the length change. The longer the column, the more interaction occurs and greater is the efficiency of the separation. Resisting this is the turbulent diffusion of the separated samples, which leads to band broadening and decreased efficiency. Shortening a capillary column to remove nonvolatile materials or plugs will decrease efficiency, but you will not notice so right away unless you start hacking off big chunks. Do it in moderation; plugs and nonvolatile compounds are trapped in the first few centimeters.

5. *Carrier gas chemistry.* The chemical nature of the carrier gas can have a dramatic effect on efficiency n of column operation. Because of its relative high viscosity, nitrogen is a poor carrier gas with a low range of useable gas velocity, and its efficiency drops off rapidly at high flow. Helium and hydrogen are both much better choices, with hydrogen the performance gas of choice. Because of its explosive nature, proper venting of hydrogen is obviously important.

6. *Carrier gas pressure.* Increasing pressure increases the retention k of the sample in the stationary phase. The sample has a longer period to interact with the column and to improve separation. In systems offering programmable electronic pressure control, this variable offers real potential as a gradient control variable in methods development.

7. *Temperature.* Temperature is the major control variable used in gas chromatography. Elevated temperature decreases retention time k, but it also can lead to separation α effects. Peak positions do not always maintain their relative position as the temperature is increased. This can be useful when the effect causes peak changes in the correct direction, but the effect is difficult to predict. Because the effect is not instantaneous, there is a lag time that varies with oven design. This leads to some variations in methods when running samples on different manufacturers' equipment. Electronic pressure control offers a simple retention factor variable with instantaneous response.

3.5   DERIVATIVES

To be successfully analyzed by GC, a compound must be volatile. But what do we do if it is not? One of the techniques used is to derivatize the compound, which often increases its solubility, especially if the compound to be analyzed is charged or highly polar. But, why would adding mass to a molecule increase its volatility?

The clue to understanding what makes this work is in the kinds of functional groups that are used in derivatization. Compounds such as bistrimethyl-silylacetamide (BSA) or bis-trimethylsilyltrifluoroacetamide (BSTFA) place trimethylsilyl or trifluoromethylsilyl groups on active hydrogen sites in amines, alcohols, or carboxylic acid groups. Alkyl ester derivatives are formed from carboxylic acid groups: oximes are made from ketones and aldehydes. When untreated, all of these groups have one thing in common: they form hydrogen bonds and they aggregate. Hydrogenbonding interaction reduces volatility. Derivatives that prevent hydrogen bonding or remove hydrogen bonding functional group increase a compound's volatility.

Crude samples that must be derivatized before analysis generally have to be extracted into an organic solvent and dried before being reacted. Derivatizing agents will react with the hydrogens in water as readily as with the target compounds. Catalysts such as trimethylchlorosilane or reagents such as pyrimidine must sometimes be added to complete the reaction. All reactants added to the mixture are potential contaminants for the mass spectral analysis and must be removed, either by GC or by preinjection extractions. If possible, avoid derivatization; it simplifies the separation.

4

THE MASS SPECTROMETER

The basic components of the mass spectrometer are the pumping system, the interface to the gas chromatograph, the ionization chamber and electron source, the focusing lens, the quadrupole analyzer, the detector, and the data/control system. Pumping systems providing high vacuum (10^{-5} Torr) are critical to the operation of the mass spectrometer. Electrons and ionized compounds cannot exist long enough to reach the detector if they suffer collisions with air molecules in the analyzer.

4.1 VACUUM PUMPS

Vacuums for mass spectrometry are established in two stages, a fore pump takes the vacuum down to 10^{-1} to 10^{-3} Torr, then either a oil diffusion or a turbo-molecular pump drops the analyzer pressure to 10^{-5} to 10^{-7} Torr (Fig. 4.1).

Vacuum is measured in either Torr or Pascal units. The Torr, equal to the pressure of 1 mmHg, is a commonly accepted measure of vacuum in the United States. The Pascal, equal to 7.5×10^{-3} Torr (mmHg), is more commonly used in Europe.

Vacuum pressures are measured by two types of gauges. The medium-level vacuum of the fore pump can be measured by a thermoconductivity gauge,

GC/MS: A Practical User's Guide, Second Edition. By Marvin C. McMaster

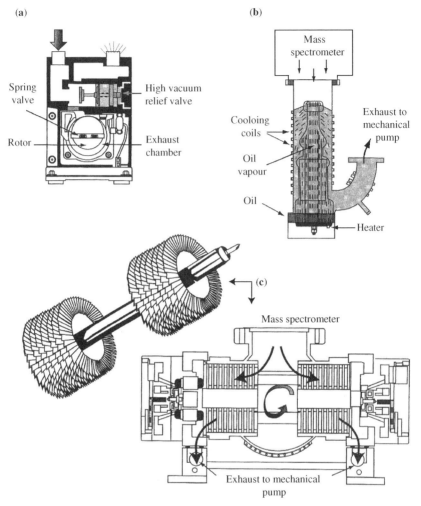

FIGURE 4.1   Mass spectrometer vacuum pumps. (**a**) Rotary-vane vacuum pump. (**b**) Oil diffusion pump. (**c**) Turbomolecular pump.

such as a Pirani gauge. A heated wire is exposed in the vacuum line and is cooled by contact with molecules. The lower the contact rate, the lower the current drawn, and the lower the vacuum reading. High vacuums produced by a oil diffusion or turbo pump require use of a hot cathode gauge. Electrons streaming from the cathode are lost through contact with air molecule. The current produced is proportional to the concentration of air molecules present.

The mechanical roughing or fore pump is an oil-sealed, rotary-vane vacuum pump commonly used as the laboratory workhorse vacuum pump. A piston on an eccentric drive shaft rotates in a compression chamber sealed by oil-lubricated spring-loaded vanes and moves gas from the inlet side to the exhaust port. It can only reach 10^{-3} Torr vacuums because of the vapor pressure of the sealing oil. Mechanical pumps typically exhibit pumping capacities of $50-150$ l/min.

The oil diffusion pump sits between the inlet port of the rough pump and the outlet of the mass spectrometer. Vacuums should be below 10^{-2} Torr before the diffusion pump heater is turned on. Heated oil rises up the pump chimney, jets out through circular opens at various level, condenses on contact with the cooled walls trapping gases from the mass spectrometer, and runs down the sides exhausting entrained gases into the roughing pump inlet. Diffusion pumps reach vacuums of 10^{-9} Torr when chilled with liquid nitrogen. They can have capacities as high as $200-500$ l/s, which can be important when pumping sources using high levels of gases for chemical ionization or when running ion spray HPLC interfaces. Many systems with oil diffusion pumps have butterfly valves in the exhaust throat that snap shut in case of power loss to prevent contamination of the analyzer with pump oil. This is an excellent feature if you are responsible for cleaning the analyzer.

The turbomolecular pump, commonly referred to as a turbo pump, is a jet engine for your mass spectrometer. It has a series of vaned blades on a shaft rotating at speeds up to 60,000 rpm between an alternate series of slotted stator places. Air is grabbed by the blades, whipped through the stator slots, and then gabbed by the next blade. Only a small amount of air is moved each time, but the number of blades and the high rotary speed rapidly move air from the analyzer chamber to the exhaust feeding into the rough pump. Most turbo pumps have a dual set of vanes and stators on a single shaft feeding a dual exhaust.

The turbo pumps on many desktop systems are only the size of your doubled fist, but they can bring the analyzer pressure down to 10^{-8} Torr on a good day. They do not have the pumping capacities of larger turbos that can move $150-2500$ l/s. They are used on systems having only electron impact (EI) interfaces that do not run high source pressures. The biggest advantage of the turbo pump is that it contains no oil to contaminate the analyzer. Its biggest drawback is mechanical failure, although that has been constantly improved on. Work with a manufacturer that has a good turbo pump trade-in program, never try to rebuild a turbo pump. That is definitely a job for a skilled professional.

It is important to vent a turbo pump to the atmosphere before turning it or an oil diffusion pump heater off. Oil vapors can be sucked into turbo pump from the rough or oil diffusion pump if the turbo pump is left under vacuum. Some systems use all three pumps: rough pump, connected to an oil diffusion pump on the source, connected to a turbo pump on the analyzer. In shutting down these systems, turn off the diffusion pump heater, allow it to cool below 100°C, vent the system, then switch off the turbo pump. These differentially pumped system are very important if you are running a chemical ionization source where you will have a very high source pressure. They also allow you to run GC effluent directly into the source without using a separator interface to reduce the volume of sample.

4.2   INTERFACES AND SOURCES

The interface between the gas chromatograph and the mass spectrometer is critical for system performance. It transfers sample from the gas chromatograph into the mass spectrometer source without mixing separated bands. It also can be designed as a separator to concentrate the sample about 50-fold and to reduce the source pressure by removing much of the carrier gas. It can be designed to exchange the carrier gas with a makeup gas to aid in running chemical ionization. That is the good news. The bad news is you are probably stuck with the interface that the manufacturer selected when optimizing your system.

The basic interface is a direct connection of the capillary column end into the sample inlet port on the ionized source of the mass spectrometer. A differentially pumped MS system with a high volume transfer pumping system on the source and a separate high vacuum system for the analyzer would be able to handle the complete GC column feed. On a mass spectrometer with a single source of high vacuum pumping, this direct connection may supply sample too rapidly. The vacuum system may not be able to maintain the 10^{-4} Torr vacuum needed for ionization. Too many ion-to-molecule collisions would occur to provide an ionized sample stream to the analyzer. Detector sensitivity would drop and signal would eventually disappear if GC flow rate was too high.

The next improvement would be to add a splitter T connection with a needle valve on the mass spectrometer side. This allows a control amount of sample to be diverted to waste or to a secondary detector (Fig. 4.2).

Improving on this, we can add a rotary-vane mechanical pump vacuum system on the exhaust of the T and add a jet separator in the capillary line from the gas chromatograph to the mass spectrometer. Sample from the capillary column expands into separator chamber. Because of its low molecular mass, carrier gas is easily diverted into the vacuum exhaust. The high molecular weight sample maintains its momentum into the MS source. There is a loss of sample mass, a much high loss of carrier gas, and a net reduction of the sample stream pressure in the source.

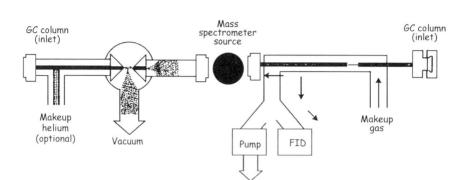

FIGURE 4.2   Mass spectrometer interfaces. (**a**) Molecular jet separator. (**b**) GC/ MS interface.

A number of sources have been designed for mass spectrometer sample ionization: electron impact (EI), chemical ionization (CI), fast atom bombardment (FAB), and field ionization (FI), with the EI source being the most common. Only the first two are commonly used in GC/MS laboratories.

The EI source (Fig. 4.3) exposes the sample from the gas chromatograph interface to a stream of 70-eV electrons from the filaments.

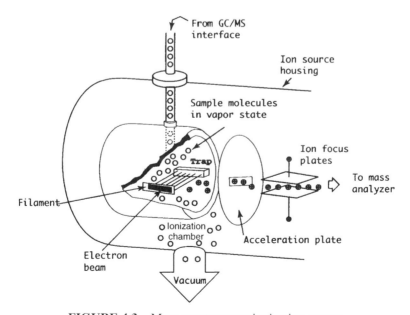

FIGURE 4.3   Mass spectrometer ionization source.

The sample molecules have an electron knocked off, or expelled, leaving behind a molecular ion with a positive charge. This ion is forced from the ionization chamber by a positively charged repeller on the back wall. The stream of ions passes through a slit, or pinhole, into a series of electrically charged focusing lens into the quadrupole mass analyzer area. The analyzer vacuum of 10^{-5} to 10^{-8} Torr helps move ions and prevents collisions with uncharged molecules or with each other.

The 70-eV energy of the impacting beam is high enough not only to ionize the sample molecule but also to cause many of them to fragment. The fragmentation pattern of the ions formed at a given electron energy is characteristic of the ionized molecules (Fig. 4.4).

Every time a molecule of the same compound is ionized under the same conditions, it forms the same quantity and pattern of ions. This fragment pattern becomes a fingerprint that can be used to identify and quantitate the molecule being analyzed. The limitation of the technique is that under the voltage used, many molecular ions first formed do not survive fragmentation. Since this molecular ion gives us the molecular weight of the compound, it is sorely missed when it is absent from the EI spectrum.

This bring us to the second commonly used mass spectrometry source, the CI source. The CI source uses an ionization gas mixed with the sample stream in an enclosed ionization chamber. Gases such as methane, butane, and carbon dioxide are used to absorb the initial ionizing electron. Since the diluent gas in present in much higher concentration than the sample molecules, its molecules have a much higher probability of being struck by the electron stream and losing an electron. Through collision, they meet and transfer energy through a chemical process to the sample molecule that is ionized, in turn freeing an uncharged diluent gas molecule. The chemical ionization of the sample molecule occurs at much lower energy than in

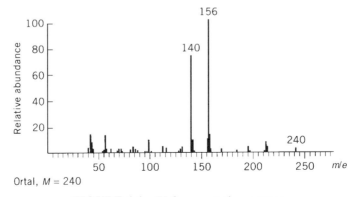

FIGURE 4.4   EI fragmentation pattern.

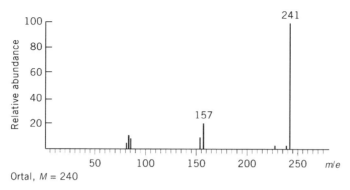

FIGURE 4.5   CI fragmentation pattern.

electron impact ionization (Fig. 4.5). The sample ion formed is more stable and usually retains the molecular ion structure without fragmentation or rearrangements.

When analyzed in the quadrupole, the molecular ion appears as a very strong, if not the major, fragment in the mass spectrum. Since it is the largest fragment present, it can be used as a quick identification of the molecular weight of the sample molecule.

Be aware, however, that there are sample preparation artifacts, such as sulfuric acid adducts from extractions, which can produce compound molecular ions with masses larger than the expected molecular weight. Another problem is that some compounds do not form stable molecular ions even under CI conditions and may only exhibit a faint molecular ion fragment.

4.3   QUADRUPOLE OPERATION

Once the sample is ionized, it and its ionization fragments must be focused, propelled into the analyzer, selected, and the number of each fragment formed must be counted in the detector.

The first step in moving the charged ion fragments into the analyzer is provided by a repeller plate at the back on the ion source equipped with a variable voltage charge of the same sign as the ionized fragments. This forces the ions through a pinhole entrance into the higher vacuum area of the analyzer. Just past the entrance hole is a series of electrical focusing lens (Fig. 4.6).

Variable voltage charges with the same polarity as the sample ions on these lens squeeze the ion beam into an intense stream as it enters the quadrupole analyzer.

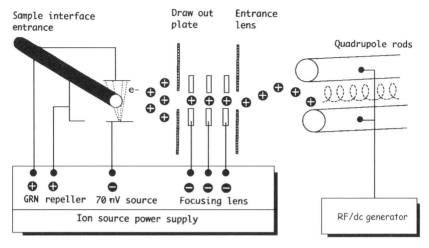

FIGURE 4.6   Focusing lens.

The quadrupole mass analyzer is the heart of the mass spectrometer. It consists of four cylindrical quartz rods clamped in a pair of ceramic collars. The exact hyperbolic spacing between diagonally opposed rods is critical for mass spectrometer operation. Rods should not be removed from the ceramic collars except by an experienced service organization.

Both a direct current (dc) and an oscillating radio frequency (RF) signal are applied across the rods with adjacent rods, having opposite charges (Fig. 4.7).

The ion stream entering the quadrupole is forced into a corkscrew, three-dimensional sine wave by the quadrupole electromagnetic field of the

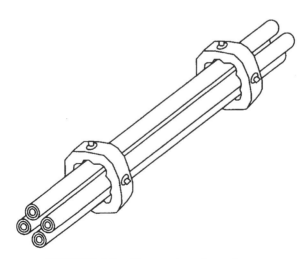

FIGURE 4.7   The quadrupole analyzer.

analyzer. The combined dc/RF field applied to the rods is swept toward higher (or lower) field strength by the dc/RF generator, upsetting this standing wave for all but a single-fragment mass at a given RF frequency. This single mass follows a stable path down the length of the analyzer and is deflected onto the surface of the ion detector. Any ion fragments not passed at a given RF frequency follow unstable decaying paths and end up colliding with the walls of the quadrupole rods. As the RF frequency is swept up or down, larger or smaller masses strike the ion detector.

4.4   THE ION DETECTOR

The fragments that pass the analyzer strike the surface of the detector after first being deflected away from a straight path out of the analyzer by a lens called the amu offset (Fig. 4.8).

Gamma particles produced in the electron ionization source are not deflected by the amu offset but cause false signals if allowed to directly impact the detector surface. The fragment ions striking the detector surface induce a cascade of ions within the detector body, amplifying the single-fragment signal sending a strong enough signal for the data system to process.

The combination of fast analyzer scanning, fast detector recovery, and high capacity data systems allows acquisition of about 25,000 data points per

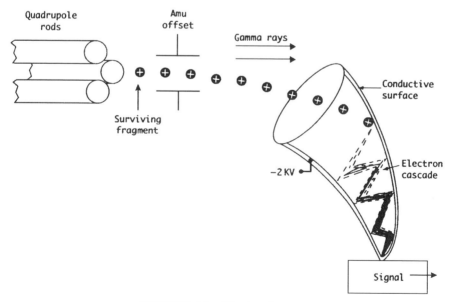

FIGURE 4.8   The ion detector.

second. This means that a mass spectra run in a SCAN mode from 35 to 550 m/z can average $8-10$ scans in 1 s. Run in a single-ion (SIM) mode, the same mass spectrometer could analyze 10 m/z mass regions in a step scan and gain a tremendous gain in sensitivity by average a much higher number of points at each of the 10 m/z locations.

5

GETTING STARTED IN GC/MS

The purpose of this chapter is to walk through the procedure of setting up and running a GC/MS system. We will describe how to make an injection, produce a scanned total-ion chromatogram (TIC), extract a spectrum, and do a library search.

We will simulate setting up a GC/MS system by doing a sample extraction, equilibrating the GC oven, getting the mass spectrometer under vacuum, and programming the run. Next we will calibrate and tune the mass spectrometer for the run. We will then run a manual injection of the sample and collect the data chromatogram. Finally, we will examine the data, extract a single-ion chromatogram (SIC), a mass spectrum, and run a library search of an online database to identify the compound.

5.1 MODE SELECTION

Before we can even start evacuating the mass spectrometer, we must make a choice of the mode of analysis we will be using. The electron ionization (EI) mode will give us fragmentation information that can be used to identify the compound. The chemical ionization (CI) mode will allow us to determine the compound's molecular weight and may give some fragmentation information.

GC/MS: A Practical User's Guide, Second Edition. By Marvin C. McMaster
Copyright © 2008 John Wiley & Sons, Inc.

Each mode requires insertion of a different ionization source before your mass spectrometer is put under vacuum. Many desktop systems may not have the option of using a CI source. They may not have an available CI source module or sufficient pumping capacity to evacuate the source chamber against the high concentration of diluent gas. Since, in this run, we plan to do a library search on fragmentation spectra, we will have to use the EI source. Most existing libraries are standardized on 70-eV EI fragmentation data from quadrupole and magnetic sector mass spectrometers.

The next mode selection we need to make is on how we will scan the mass spectrometer. We can choose to do a continuous scan over a range of masses (SCAN mode) or we can a jump scan over a discrete number of masses (SIM mode). It depends on whether we wish to look at the forest or at the trees. We can chose to see a broad overview view (SCAN) of the forest or to look at specific trees in that forest with very high sensitivity (SIM). For unknown or complex mixtures, we almost always run SCAN, at least for a first run. When we are looking for trace amounts of specific compounds or looking for changes in composition over a very long period of time, we would choose SIM.

Here we have chosen to run a scan from 50 to 550 amu. We select scans above 50 amu to avoid traces of water (18 amu), nitrogen (34 amu), and oxygen (36 amu) from any air residues, although me might look for these in our system tuning and performance evaluation. We also will probably be scanning from high mass to lower mass, because we get less tailing and, therefore, better resolution between any mass peak M and its carbon isotopic peak $M + 1$.

5.2 SETTING UP

Before making an injection, the capillary column must be connected to the injector and the mass spectrometer; the mass spectrometer must be at an acceptable vacuum, a clean appropriate throat liner must be in the injector, and the column oven must be at the equilibration temperature and temperature programmed for the run. The sample must be prepared for injection. The mass spectrometer must be calibrated and the scan range for the run programmed, before we may proceed. In most cases, once the daily autotune has been run, the only thing you need to do is program the GC oven and set the mass range before making the injection.

The column we will use for the injection is a DB5 capillary column containing a bonded phase of 5% phenyl silicone and 95% methyl silicone. Connect the column head through the ferrule into the injector body, slide the fitting and ferrule over the column tail, and connect it to the mass spectrometer interface.

Once the source body, ionization filament, and lens are inserted into the quadrupole mass spectrometer, we are ready to begin evacuating the mass detector. In most laboratories, the mass spectrometer is usually kept under vacuum at all times unless it is down for cleaning.

To start up a mass spectrometer, you first turn on the mechanic oil pump and pump until you have reached a vacuum of 10^{-3} to 10^{-4} Torr. For an oil diffusion pump, turn on the oil heater element and begin jacket coolant circulation. If the high vacuum pump is a turbomolecular pump, it can be switched on at this point. It should be noted that when you are reversing the process and are shutting down, the turbo pump should be vented before you shut off the mechanical vacuum pump to prevent oil from going back into the turbo pump. When the pump pressure reaches 10^{-6} Torr, you are ready to begin calibration. Do not be in a hurry since this might take 4 h or longer.

A clean, tight system with high capacity pumping may reach 10^{-7} Torr; if vacuum fails to go below 10^{-5} Torr, start checking for leaks. When you start your evacuation, listen for a change in the sound of the rotary-vane pump. If you do not hear it within 10 min, push down on the lid of the vacuum containment system to make sure the gasket is sealed. Once vacuum has been achieved, turn on the filament and scan from 0 to 50 amu to see if you are getting water and air peaks. If so, check for leaks in the system.

Sample preparation may be as simple as dissolving a sample of your mixture in a solvent. Or you may have to first extract your sample from an acidified aqueous phase, dry, evaporate, and derivatize it. For mixtures of standards or unknowns, you will probably add at least one internal standard, to correct for injection and retention variations, and possibly a surrogate standard, to correct for sample recovery during extraction. You also may run sample blanks and extraction blanks.

If you do not know the history of the last run on the column, you may want to run a quick bake out before setting equilibration. Turn on your carrier gas, run from 50 to 300 at 30°C/min, hold for 2 min, then do a step down to the equilibration temperature. Check your column specification for the maximum purge temperature for your column, especially if you are using a nonbonded phase column. Overheating can cause excessive bleeding and separation from the support.

When your sample is ready for injection, turn on the gas chromatograph and set the equilibration temperature for the injection. This is usually around 50–120°C, but we will use 50°C for this injection. We can now program the column oven for our run. We will set a hold at 50°C for 1 min, and then ramp from 50 to 320°C at 30°C/min.

Since we decided we would use SCAN mode for our run, we need to select a scan range of 50–550 amu. We will work above 32 amu to avoid contamination from dissolved air. The mass spectrometer filament needs to be protected from the slug of methanol from the injection rushing down the capillary column. If you had one, you could set the auxiliary valve 1 after the GC column outlet to open from 0 to 2 min to divert solvent away from the mass spectrometer. In our case, we will use a time program to simply avoid turning on the mass spectrometer filament until after the methanol bolus has passed though the source chamber.

5.3   MASS SPECTROMETER TUNING AND CALIBRATION

Before a mass spectrometer can be used to measure masses of fragmentation ions, it first must be calibrated and tuned. Calibration means adjusting the dc/RF signal frequency so that each mass axis point corresponds to the expected mass fragment position from a calibration compound. Tuning is done using lens adjustments to insure that adjacent mass peaks overlap as little as possible and that relative peak heights for the tuning compound fragments have the expected ratios along the voltage axis. Calibration can be looked on as making adjustment along the mass (X) axis while tuning adjusts the relative peak intensities along the voltage (Y) axis. Calibration and tuning are done so that the same compound, run on different machines under the same operating conditions, will always exhibit the same fragment masses in the same relative amounts. In many machines, a single tool called autotune can adjust both calibration and tuning. It usually provides an adequate calibration, but additional tuning is usually required for separation of complex mixtures.

Calibration is done with a volatile liquid called perfluoro-t-butylamine (PFTBA or FC43) referred to as calibration gas (Fig. 5.1). This "cal gas" is placed in a bulb valved off from the mass spectrometer source. When it is needed, the valve is opened and some of the PFTBA volatilizes into the source chamber and is ionized. The fragmentation pattern produced has characteristic bands at 69, 131, 219, 264, 414, and 502 amu that are used to adjust the mass axis. Generally, adjustments are made first on the 69 peak, which should be the largest in fragmentation pattern, and then you work toward the 502 peak, which is the smallest. Adjustment of the repeller and tuning lens setting by hand are used to move the 69, 131, and 219 mass peaks to the correct positions. If you get the 69 peak correct, the others begin to drop into position. The 502 peak is very small and the hardest to find. Adjustment of the amu offset lens and the ion detector voltage will generally increase its size to a point where it can be seen.

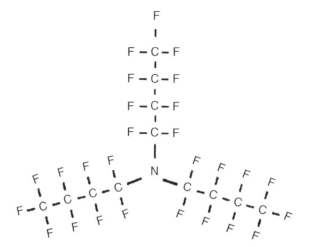

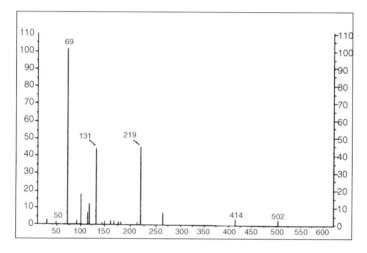

Mass	% relative abundance
50	1.2
69	100
131	45
219	45
414	1.9
502	0.9

FIGURE 5.1 PFTBA calibration compound.

For our first run, we will turn on the cal gas, then run autotune, and let the autotune software adjust the lens voltages for us. When it is done, we will inspect the spectrum produced. Mass 69 should be the major or base peak, 131 and 219 should be about equal and a little over one-half as large as the 69 mass, and 502 would be at around 2−5% of the 69 peak. We will discuss variables that control peak heights, lens adjustments to vary them, and tuning compounds in Chapter 7.

5.4 SAMPLE INJECTION AND CHROMATOGRAPHIC SEPARATION

Our sample for this injection will be a mixture of four phthalate ester standards dissolved in methanol. We will make 100× solutions of 500 µg/ml each of dimethyl, di-*n*-butyl, benzylbutyl, and di-*n*-octylphthalate. Next we will add a 1-ml sample of each compound to a 100-ml graduated cylinder and dilute with methanol to make a 5-ng/µl injection sample. The 100× sample should be centrifuged or filtered through a 0.54-µm filter. This same mixture will be used later in Chapter 7 as a column quality control standard to study column performance.

We will inject 5-µl of our 1× sample into the injector using a splitless injection with injector temperature set at 275°C and 5 cm/s helium purge activation at 45 s. Both column oven program activation and mass spectrometer scan start are triggered by the injection of the sample. After the delay programmed to allow the injection solvent to pass through the mass spectrometer source, the filament in the MS source will switch on, and we will begin to see the TIC of our calibration test mixture on the data system. Figure 5.2 is a chromatogram of a commercially available mixture that includes these four phthalates.

The first peak, methylphthalate, should come off after about 7 min and the fourth peak, di-*n*-octylphthalate, should elute after 10 min. The retention time and efficiency plate count of these two compounds should be calculated and stored for reference at a later time when we feel that the column has changed separating character. We can pull the standards out of the freezer and rerun them as a independent check of column performance.

5.5 DATA COLLECTION PROCESSING

As we discussed in Chapter 1, the data system stores mass spectrometer data as a three-dimensional block with three axis: time, intensity, and

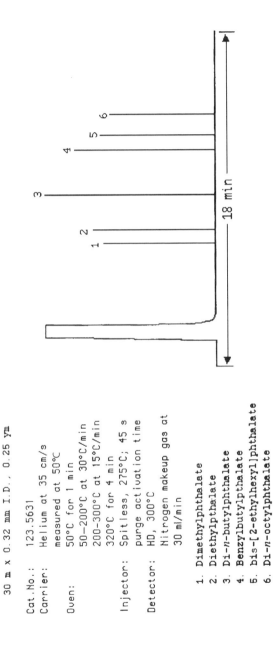

FIGURE 5.2 Calibration test mixture.

Phthalates—column: DB" -5.625

30 m x 0.32 mm I.D., 0.25 ym

Cat.No.: 123.5631
Carrier: Helium at 35 cm/s
measured at 50°C
Oven: 50°C for 1 min
50–200°C at 30°C/min
200–300°C at 15°C/min
320°C for 4 min
Injector: Spitless, 275°C; 45 s
purge activation time
Detector: HD, 300°C
Nitrogen makeup gas at
30 ml/min

1. Dimethylphthalate
2. Diethylpthalate
3. Di-*n*-butylphthalate
4. Benzylbutylphthalate
5. bis-[2-ethylhexyl]phthalate
6. Di-*n*-octylphthalate

18 min

mass/charge (*m/z*), see Fig. 1.6. The TIC in Fig. 1.5, is a summation of the intensities of all mass fragments at a given time and is only one way of displaying the data in two dimensions. We can also make planar slices of the data and display them.

If we make our cut at a given *m/z*, we will display a SIC, intensity versus time, see Fig. 1.7. At first glance, the SIC appears to be similar to the TIC, but with fewer peaks. This is because the SIC is only displaying compounds

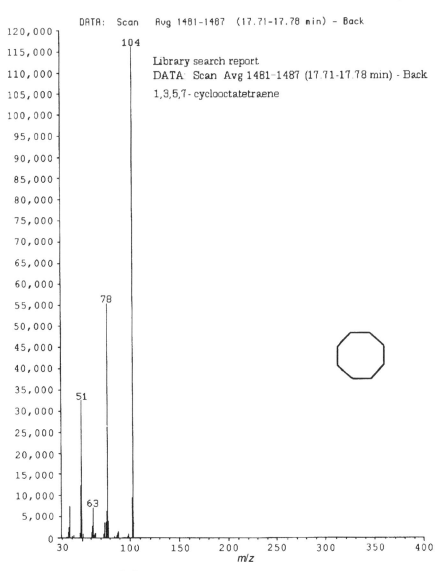

FIGURE 5.3   Spectral library report.

containing a single mass fragment. If a compound does not contain that fragment, it will not appear in the SIC. The SIC is very useful for determining related compounds that break down to form common intermediates.

If we make a cut at a given time point, we can display this as a mass spectrum, intensity versus *m/z*, see Fig. 1.8. The spectrum presents all the fragments associated with the chromatographic compound present at the selected time point. With these data, we can determine the structure of the compound using mass spectral interpretation techniques. We can also use the data to identify a known compound by comparing it to a spectral library (Fig. 5.3).

Library searching is done using probability matching comparison of known and unknown spectra, usually starting with the largest peaks present and then working down toward smaller peaks. The hit list contains all the possible matching spectra and the matching probability for each spectra starting with the match of highest probability. Library software will usually display the spectrum, the best matching spectrum that was found, a difference spectrum, and compound information for the matching compound, such as structure, molecular weight, other compound names, and physical data for the compound.

PART II
A GC/MS OPTIMIZATION

6

CHROMATOGRAPHIC METHODS DEVELOPMENT

Methods development in GC/MS focuses mainly on one variable, oven temperature control. The second variable used in GC methods development, column chemistry change, has a lessened impact on GC/MS because of the necessity to break and then restore the MS vacuum in order to insert a new column. Column separation changes are primarily due to changes in the polarity of the packing materials in various columns. Nonpolar packings more strongly attract nonpolar compounds that elute later than polar compounds do. Hydrophilic, polar packings have more of an affinity for polar compounds, with nonpolars eluting more rapidly.

Carrier gas flow rate and viscosity affect sample residence time on the column and the column pressure. The faster the gas flows down the column, the shorter the sample residence time and the less chance compounds have to separate. Flow rate control offers some potential for modifying a separation since it can be changed continuously. The limiting factor seems to be rapid fall of separation efficiency with increases and decreases in flow rate changes, beyond an optimum flow rate.

Changing an injector throat liner can also affect a GC separation by increasing or decreasing the contact surface area, changing the character of the vaporized sample that actually enters the column. However, change cannot be made in a stepwise manner, and its effect is usually not predictable.

GC/MS: A Practical User's Guide, Second Edition. By Marvin C. McMaster
Copyright © 2008 John Wiley & Sons, Inc.

Controlling the split ratio of a split–splitless injector can also affect the content of volatile material reaching the column head. But since it is controlled by the volatility of each component, the temperature of the liner, and the split residence time, it is very difficult to predict the exact composition of the injected sample. In actual practice, both these techniques are changed based on empirically derived information and, while their affects can sometimes be dramatic, they are at best unpredictable.

6.1   ISOTHERMAL OPERATION

In isothermal operation, the column temperature is set one time and the column is allowed to reach a constant temperature before the sample is shot into the injector. Once the sample is vaporized in the injector and swept on to the column, it must interact with the column coating. This is often aided by running the column head at a lower temperature than that of the injector throat, making the sample partially condensed and concentrated. The interaction of the sample with the packing material also aids this concentration, leading to a sharpening of the disk of sample within the column. An equilibration is established between the concentration of each component in the coating on the packing and in the vapor phase above it.

This equilibrium is continuously upset in favor of the vapor phase component when the carrier gas sweeps down the column. As multiple interactions occur down the length of the capillary column, components with lower interaction affinity for the coating move more rapidly and begin to separate from the more highly retained material. Concentration disks of individual compounds begin to separate. As they travel down the column, diffusion effects, packing voids, and wall interactions begin to distort the shape of the separation disk. The disk broadens, the center tends to move faster than the outside edges, and it is pulled a bit into the shape of a nosecone. Finally, they reach the detector interface, enter the detector cell, and appear in the data system as a sharp front peak with maximum concentration at the peak center and some tailing on the backside as the trailing edges of the "nosecone" emerge. The longer a peak stays on the column, the broader its disk will become due to diffusion, but the more chance it will have to separate from other components of similar affinity for the packing.

Unless all components have a similar volatility at the column temperature, the later-eluting components will begin to show broader and broader peaks, with more and more tailing, until the last components flatten down to the baseline. The separation still occurs, but not in a useful period of time. To speed this separation, we could equilibrate the column at a higher initial temperature. This, unfortunately, leads to a compression of the separation we achieved with the early-eluting components. Fast-moving compounds do not have enough time on the column to fully separate. In the worst case, they will coelute as a broad, unresolved peak at the separation front.

6.2 LINEAR TEMPERATURE GRADIENTS

To resolve a mixture of compounds with widely differing retention on the packing, you need to run a linear temperature gradient. You equilibrate the column at a temperature that resolves early running peaks and then gradually increase the temperature to a final temperature that will remove all the components of interest. Hold at the final temperature long enough to resolve the last two peak, and then drop the oven temperature back to the starting temperature for the next injection. It is important to hold the temperature at this point long enough to equilibrate the oven. Failure to do so will cause disturbances in the first part of the chromatogram; too low a temperature will cause early-eluting compounds to retain longer. The more common problem is that the oven temperature will still be too high when you make the next injection and early runners will elute too early and be jammed together. If peaks are spread in the middle of the chromatogram, use a faster temperature rate increase to draw them together, although you may find this jams late-running peaks together.

6.3 ASSISTED RE-EQUILIBRATION

When you are trying to make as many injections as possible in the golden time between mass spectrometer tunings, eliminating dead times is critical to your success and profit margin. One of the major dead periods is the time required to cool and re-equilibrate the GC oven.

The first method used to overcome this problem is to activate a piston used to push open the oven door at the end of the temperature ramp. The piston holds the door open until the starting temperature is reached and a spring pulls the door shut when the piston retracts. The next modification uses air blasting of the oven combined with the door opening to further accelerate the temperature drop. In the final method, cryogenic cooling from adiabatic expansion of compressed helium is added to further accelerate the cooling process on some machines.

Care must be taken that the cooling does not become too vigorous resulting in extended heating time to reach the starting temperature. The re-equilibration time is usually fast enough not to require further modification.

6.4   HINGE POINT GRADIENT MODIFICATION

Next, we must deal with compressed and expanded areas in the chromatogram. Compressed areas are sections of the separation where poor resolution causes many peaks to overlap. In an expanded area, peaks are over-resolved and are increasing the total separation time. For each of these, we can identify a "hinge point" before the first peak of the area (Fig. 6.1a).

Both can be handled by altering the rate of temperature change at these hinge points. The temperature change must be made while the sample is still in the column to have an effect on the elution pattern.

The first step in this hinge point development is to run a linear gradient and optimize the separation of early and late peaks using proper equilibration and final hold temperatures and times. Inspect the first compressed area in the chromatogram and determine the retention time of its first peak (t_{c1}). Measure the void time of the column by measuring the time from injection to the first peak front (t_0) for unretained solvent in the chromatogram (Fig. 6.1b).

Measure back from the start of the first peak in the compressed area an amount equal to the void time of the column ($t_{c1} - t_0$). Go to the column

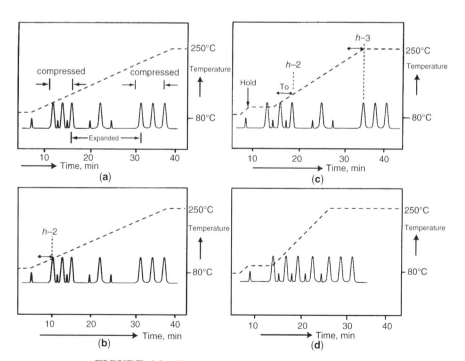

FIGURE 6.1   Temperature gradient programming.

temperature profile program and enter a temperature hold starting at this time point with duration equal to the time width of the compressed area in the linear gradient. After this hold, enter a program step to return to the original temperature ramp rate. Reinject the sample and look at the effect this change produced (Fig. 6.1c). Repeat the process for each compressed area in the separation.

For an expanded area of peaks, determine the hinge point, drop back the equivalent of the void time, and double the ramp rate over a time period equal to one-half the compressed area. Then, enter a programming step returning to the ramp rate and reinject (Fig. 6.1d). Handle other expanded areas in the same fashion until you have achieved the separation you want.

If you find that a temperature hold spreads the compressed separation too much, go back and replace the hold with a ramp with a slope half way between the original ramp rate and the hold for the same duration of time. If doubling the ramp rate over a compressed area causes too much compression, pick a rate half way between the old and new ramp rates and repeat the injection.

Remember oven temperatures turn like an 18-wheeler, not a sports car. They have to have time to produce an effect. Peltier-type electrical radiative heating and cooling of a much more confined column heater space could offer more precise control of both heating and re-equilibration. It also would provide more rapid, precise temperature changes for developing fast, complex temperature gradients. Commercial GC systems using this type of column temperature control are now coming on the market using this type of heating, especially where compact gas chromatograph are needed such as in the space program and in man-carried GC/MS systems.

6.5   PRESSURE GRADIENT DEVELOPMENT

Electronic (column) pressure control (EPC) is a relatively new technique for making controlled, predictable separation changes. It was designed to correct for pressure drops occurring during injection that led to variations in sample run times due to variations in sample size. Since pressure can be varied continuously and can produce changes in the time a sample stays on the column, EPC is very useful as a control variable. Increase the pressure and decrease the residence time. Decrease the pressure and increase the time the sample has to interact with the packing materials and become separated.

Unlike column temperature control, which has lag time associated with heating and cooling the large volume of the oven, programmable pressure control produces a nearly instantaneous effect on a separation. The limitations of this technique are how wide a pressure range the capillary column will tolerate, the accuracy of the pressure control apparatus, and the linearity of the effect of pressure change on retention of compounds on the column. It certainly should be explored as a retention time variable for methods development and separation control if this tool is available on your GC system.

6.6   COLUMN REPLACEMENT

Another area that now can and should be explored for method development is rapid change of column coating types to produce an alpha effect on separation. Changes in column chemistry can make a dramatic change in separations, even change the order in which peaks occur, aiding in building methods. Presently, to change a GC column, it is necessary to break the mass spectrometer's vacuum and to shut the system down before inserting a different column. Because of the long delay in re-establishing the operating vacuum, column change is seldom used.

Agilent has introduced a new product called QuickSwap (Fig. 6.2), a microfluidic device where a constant pressure source of makeup gas purges the column connection point when the column is removed. The device was designed to be fully integrated with EPC control. The question remains as to how easy QuickSwap can be integrated into and controlled by other GC/MS systems.

Rapid column changes would also be useful in troubleshooting system problems. Probably 70% of all system problems involve either injector or

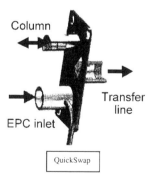

FIGURE 6.2   QuickSwap microfluidic switch. Copyright November 2007 Agilent Technologies, Inc. Reproduced with permission. Agilent Technologies, Inc. makes no warranty as to the accuracy or completeness of the foregoing materials and hereby disclaims any responsibility therefore.

column contamination. Replacing the injector liner and the septum is easy and done rapidly. Ideally, replacing the column with a blank capillary column would allow us a "column bridge" to quickly separate between column and hardware problems and to do rapid system diagnostics.

QuickSwap or a similar product offers tremendous time saving to a production facility. A contaminated column should not require a 4-h system shutdown for replacement when you are working in the golden 12-h sample running time between EPA required tunings.

7

MASS SPECTROMETER SETUP AND OPERATION

It would be nice if the mass spectrometer's mass axis came already precalibrated. Instead when the control system sends a specific dc/RF value to the analyzer, we have no way of knowing the m/z value of the selected ion fragment.

7.1 MASS SPECTROMETER CALIBRATION WITH CALIBRATION GASES

First, we must analyze the fragments from a known calibration compound, and then adjust the mass axis so that it agrees with the expected fragment mass assignments. Periodically, we must go back and check to see that analyzer contaminations or degraded electronic components have not changed the selected mass positions. If the positions have moved, then we must recalibrate.

In order to obtain repeatable analysis from instrument to instrument or from laboratory to laboratory, we must also tune our instrument. We do this by adjusting various electrical components, such as the repeller, lens, and electron multiplier voltages, so that the mass spectrometer will give the expected relative ratios of ion fragment intensities for a target compound.

GC/MS: A Practical User's Guide, Second Edition. By Marvin C. McMaster
Copyright © 2008 John Wiley & Sons, Inc.

Once target heights are tuned in this fashion, two instruments should provide the same analysis for the same sample, no matter where the instruments are located.

Put another way, calibration adjusts the x-axis of the spectrum so that we analyze the correct mass fragment at any point in a spectral scan. Tuning adjusts the analyzer so we see displayed on the y-axis the same relative height, or abundance, of ion fragments each time we run the target compound. If it works for a selected known compound, it should work with a similar unknown compound within a specified concentration range. Each type of analysis has an expected target compound or compounds and an effective concentration range for each identified compound in a mixture built into the description of the analysis.

Perfluorotributylamine (PFTBA or FC43) is a clear, volatile liquid under atmospheric conditions used as the most common calibration gas (cal gas) in mass spectrometer analysis. It is kept in a vial valved off the sample inlet or the DIP probe port. When the instrument needs to be calibrated, this cal gas valve is opened and calibration gas is allowed to vaporize into the source chamber. Cal gas is ionized and fragmented by the electron beam from the filament and passed into the analyzer where its fragments are separated and detected.

The major ion fragment masses for PFTBA are 69, 131, 219, 264, 414, 464, 502, and 614. In a well-tuned mass spectrometer, the 69 mass is the base mass; fragments 131 and 219 have approximately the same heights equal to $45-60\%$ of the 69 peak; the 414 peak is about $3-6\%$ of the 69 peak; and 502 will be 3% or less than the 69 peak height (Fig. 5.1).

PFTBA has been the predominant calibration gas used in mass spectrometry because of the mass range of its fragments, its evenly spaced major fragment masses, and the volatility of the gas under the analyzer vacuum. Early quadrupole analyzers usually had a mass range of $0-800$ amu. A few could reach 1000 amu and some research instruments offered an extended range of $0-2000$ amu. These extended ranges have become important in analyzing polymeric and multicharged molecules, such as peptides, proteins, and DNA fragments, with electrospray ionization. These extended mass ranges require calibration gases with larger mass fragments.

Perfluorotripentylamine and perfluortrihexylamine offer fragments in the $500-600$ and the 700 amu range, respectively. Perfluorokerosene offers fragments well above 1000 amu. Perfluorophenanthrene is sold as a single, liquid cal gas offering evenly spaced mass fragments from 50 to 650 amu. The larger calibration gases are not as volatile as PFTBA and may require heating of the holding vial for vaporization.

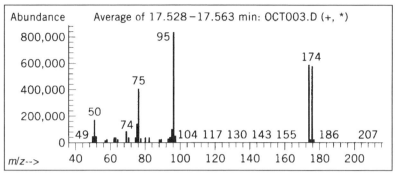

Peak apex is scan: 903

Target mass	Relation to mass	Lower limit, %	Upper limit, %	Relative abundance, %	Raw abundance, %	Result, pass/fail
50	95	15	40	20.9	178,688	Pass
75	95	30	60	48.1	410,304	Pass
95	95	100	100	100.0	853,227	Pass
96	95	5	9	6.7	57,067	Pass
173	174	0	2	0.0	0	Pass
174	95	50	100	69.7	594,304	Pass
175	174	5	9	5.8	34,216	Pass
176	174	95	101	98.5	585,557	Pass
177	176	5	9	6.6	38,552	Pass

FIGURE 7.1   BFB tuning spectra and report.

7.2   MASS AXIS TUNING

Modern high performance control systems have autotune systems that both calibrate and target tune for specific target compounds. Their performance is so good that they almost never fail to reach the desired ratios. However, not everyone has updated equipment possessing this desirable feature, and some analyses require target tuning of compounds not listed in the control system. When in doubt, run one of the available built-in target tunes first, then try the tune compound of your choice to see if it also produces the correct fragment height ratios. If not, you are going to have to tweak your tune by manual adjusting lens values or by using a lens scanner.

Autocalibration is the height of simplicity. To calibrate a mass spectrometer, set your mass spectrometer up for SCAN across the desired or specified mass range. You open the cal gas valve and push the autotune button. The system will be busy for some time while it works its way toward the final autotune. Then it will display the resulting calibration spectra on the data display. You can usually select to print a calibration report that will include the spectrum and the numerical relative peak heights of each detected mass fragment.

Target tuning is autocalibration run to produce a specific compound's calibration. The calibration gas valve is turned off and the autotune button is pushed. Using a different mathematical algorithm, it produces the desired fragmentation ratios when the injected target compound passes from the gas chromatograph through the mass spectrometer's analyzer.

The mass spectrometer is set up in a SCAN mode, the mass range for the analysis is set, the autotune button is pushed, and a specified amount of the target compound is injected into the gas chromatograph. A tune report for the target compound should display the expected ratios of large and small peak masses. Usually, this report is in the form of a pass/fail report for each fragment pair. If only one fragment ratio fails to pass, calibration lens parameters must be adjusted, and the tuning compound reinjected until a pass is achieved on all specified peak ratios. Once this is achieved, the mass spectrometer is then certified for use for a specified period of time.

At the end of this time period, tuning compound must be reinjected and the certification report must pass again. If it fails, then the target tune calibration conditions must be readjusted, followed by injection of the tune compound until the system is recertified.

To manually calibrate a spectrum, set up a calibration scan from high to low mass and locate the 69 mass, which should be the largest mass present. Adjust it with the repeller voltage until its position is at 69. Check its separation from the smaller 70 mass, there should be little if any overlap. Now adjust the mass positions of the next two largest peaks, 131 and 219, using the focusing lens until they are approximately equal in heights, about one-half of the 69 peak height. If they are much less than one-half of the 69 mass, increase the repeller setting until they are about equal to 50% of 69. Now you should be able to adjust the mass position of 414 peak relative to that of 219. If you have problems finding the 414, go back to the 264 peak and calibrate it against 219 first. You usually will have no problem finding the 414 and 502 peaks after making that adjustment. Bring the electron multiplier (EM) voltage value up until the 414 peak height is about 4% of the 69 peak. You should now be able to see the 502 peak. Adjust its mass position. You can tweak its height with the amu offset (X-ray lens) or by increasing the EM voltage setting. For calibration, the 502 peak should be between 1% and 3% of 69. For use with most tuning compounds, it will be closer to 1%.

The 502 peak height is very susceptible to source contamination and is a good measure of when to clean the source. If the other peak heights are falling in the correct ratios, but you cannot see the 502, you are probably due for a cleaning.

If you have to crank the EM voltage above 3500 V without finding 502 peak, you may also be in need of a detector replacement. I was told by a service man that the diagnostic test for the detector was to set the repeller to the maximum (10.0 V on a Hewlett-Packard mass spectrometer), open the cal gas valve, and look at the 502 peak. On a new detector, 502 should be 10% of the 69 peak height. If the value was below 5%, the detector should be replaced. He did not specify the EM voltage, but once you exceed a value of 3000 V, the detector tends to degrade very rapidly. Keep the EM voltage only as high as necessary to see the 502 peak.

In a modern instrument without autotune capability, autocalibration is run first. Then, it may be tweaked manually by adjusting various lens to achieve a specific purpose. For instance, you might use this technique in target tuning. You run your target calibration, shoot your target compound, and find that a couple of your ratios are failing. You go back to the target spectrum and see a couple of fragments are too high. You turn on cal gas, adjust your focusing lenses, turn off cal gas, and reinject the target compound and see if the tuning report passes.

Older instruments, using second-generation software, often require considerable tuning expertise. Modern instrument systems are much better, especially when the source is clean and the detector is new. In an operational environment, even they require a little tweaking. Experienced GC/MS operators earn their keep through the speed in which they can recertify a mass spectrometer's tune. Once the instrument is tuned, the operator is allowed to make as many runs as possible before the time for the next recertification. Run time equals increased analysis that translates to dollars for the laboratory.

7.3   SYSTEM TUNING FOR ENVIRONMENTAL ANALYSIS

The mainstay GC/MS analyses in the environmental laboratory are volatile organic analysis (VOA) and semivolatile organic analysis (semi-VOA). Both have tuning compounds defined that allow laboratories across the nation to report results that are reproducible. These analyses were developed by the US Environmental Protection Agency for its contract laboratories program (CLP) and picked up as standard procedures by other laboratories doing public contract analysis.

In using these compounds, the mass spectrometer is first calibrated to a specific set of tuning parameters with the tuning compound injected through the gas chromatograph. The spectra of the calibration compound are determined and the height ratios of specified mass peaks are determined.

All peak ratios must agree before the chromatographer can proceed with the analysis of standards and unknown samples. Failure of a single specified ratio will require the analyst to return to calibration with PFTBA and retune the instrument, then repeat injection of the tuning compound until agreement is achieved.

After all ratios pass, the instrument is certified for performance for a period of time specified by the method, usually 12 h. At the end of this time, the tuning compound must be reinjected and agreement of the tuning ratios verified again. Failing this, the instrument must be recalibrated with PFTBA until it can again pass a tuning report.

To run a bromofluorobenzene (BFB) tune for volatile organics analysis, set your GC oven to 230°C. Open the cal gas valve and run a PFTBA target tune or ramp your entrance lens so that 131 and 219 are nearly equal, but slightly favoring 131. Ideal conditions are 69 (100%), 131 (35%), 219 (30%), 414 (1–2%), and 502 (0.8%). Tweak the last two masses with your amu offset and the EM emission. Save your tune parameters. Close the cal gas valve.

Shoot 50 ng of BFB solution. You should get a single peak at about 17.5 min on the chromatographic display. Figure 7.1 shows a spectrum and a pass/fail tune report for BFB.

The EPA's Contract Laboratory Program (CLP) procedure requires that you select scans within 10% of the peak maximum. Other laboratory methods allow you to select any scans within the peak to pass BFB tune. Critical target mass ratios are 50/95, 75/95, 174/95, and 174/175/176/177 quadruplet. The latter are particularly sensitive to high mass variations in the PFTBA tune.

Semivolatile organics analysis uses decafluorotriphenylphosphine (DFTPP) as its tuning compound. The GC oven is heated to 250°C. The cal gas valve is opened and DFTPP target tune is run on the calibration gas. If your mass spectrometer cannot do a target tune, ramp your entrance lens until the PFTBA 131 and 219 masses are approximately equal, then tweak the amu offset and EM voltage settings to reach a 502 value of about 1%. Ideal conditions for DFTPP tuning are 69, 100%; 131, 35%; 219, 30%; 414, 1–2%; and 502, 0.8%.

Once you have a satisfactory calibration, shoot 50 ng of the DFTPP solution. You should see a single peak at about 9.4 min. Figure 7.2 shows the spectrum and pass/fail report for a successful DFTPP tune.

The base peak for the analysis is 198. Critical target mass ratios are 51/198, 125/198, 197/198/199 triplet, 442/198, and 442/443 doublet. The last two peaks are particularly sensitive to high mass variations in the PFTBA tune, to source cleanliness, and to detector aging.

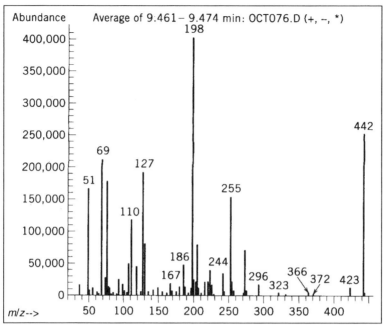

Peak apex is scan: 985
Average of three scans: 984, 985, 986 minus background scan 979

Target mass	Comparison mass	Lower limit, %	Upper limit, %	Relative abundance, %	Result, pass/fail
51	198	30	60	41.7	Pass
68	69	0	2	0.0	Pass
69	198	0	100	52.4	Pass
70	69	0	2	0.0	Pass
127	198	40	60	47.9	Pass
197	198	0	1	0.0	Pass
198	198	100	100	100.0	Pass
199	198	5	9	7.0	Pass
275	198	10	30	18.2	Pass
365	198	1	100	1.9	Pass
441	443	0	100	0.0	Pass
442	198	40	100	63.2	Pass
443	442	17	23	19.7	Pass

FIGURE 7.2   DFTPP tuning report.

7.4   ACQUIRING INFORMATION

The computer system for the GC/MS system has dual functions. We have already discussed its use in controlling the programming of the gas chromatography, the autosampler, and the mass spectrometer. Its second function is to acquire data from the mass spectrometer, process it, and generate chromatograms, spectra, and data reports from the information.

The first decision you must make in setting up the mass spectrometer for data acquisition is to select the operating modes. Most work is done in the EI mode using SCAN.

The first decision must be the ionization mode. In most systems, changing from EI to chemical ionization (CI) mode involves dropping the mass spectrometer's temperature, breaking the system vacuum, disassembling the analyzer, replacing the EI source with a CI source, reassembling the analyzer, and reevacuating the system. This may take $4-5$ h and is, therefore, not a trivial change to make. Newer systems and ion traps claim to have the ability to make this changeover without dropping the vacuum. Many laboratories will dedicate a GC/MS system to CI mode molecular weight analysis to avoid having to go through this long conversion.

The change from SCAN (mass axis scanning) to single-ion monitoring (SIM) mode is not as physically challenging, but is important because of the difference in the data produced. The SCAN mode is used when it is necessary to look at a large number of compounds in a single sample. It is the preferred mode when first examining a new sample or in doing methods development. You scan over the selected mass range at a scanning rate of about 30,000 points/s and, then average the data obtained for each mass point. Although this appears to be a continuous scan, you are actually stepping from point to point. Step, settle, measure, step to the next point, and repeat. At the end of the scan range, you must make a big step back to the start of the scan range, before starting the process again. For a mass range of $50-550$ amu, measuring a point every 0.01 amu will yield an average of six scans per second.

The SIM mode is chosen to look at a specific number of masses instead of every mass point within a given mass range. To look at four SIM peaks, you may only need look at 40 points/scan. Using the same scan rate of 30,000 points/s, you will average each point over 700 times in a second. You are measuring each point more accurately, and this translates to an increase in sensitivity. The major use for the SIM mode, therefore, is to measure a limited number of masses at very high sensitivity in trace compound analysis. The second use would be to measure a limited number of peaks over a very long time (days or months) in order to decrease the number of stored data points. While monitoring changes in an exhaust gas stream, a customer scanned for 5 s once ever hour for 5 months.

Once ionization and scanning modes are set, we are ready to begin acquiring data. The mass spectrometer is calibrated and tuned as described in the last chapter. Programs written for autosampler and gas chromatograph control are downloaded or contact closures are sent out to begin remote programs in the individual modules. An injection signal is sent either from a manual injector or from the autosampler to start the process. The gas chromatograph's oven temperature program begins to run and the mass spectrometer begins its scan program. A signal is sent to the A/D board for it to begin acquiring analog data from the mass spectrometer's detector and convert it to digital data that are stored in a designated file on the computer's hard drive. For each acquisition time point, the entire averaged mass axis data scan points must be stored. If a data point is taken every millisecond, this represents a very large volume of data. A single 40-min GC/MS data set might occupy 2–3 MB of hard drive space.

7.5   DATA DISPLAYS AND LIBRARY SEARCHES

From these stored data, information is extracted for real-time displays. For each time point, the signal intensity of each mass point can be summed to give a total ion current. Display of the ion current at all the time points yields a total-ion chromatogram (TIC). You could also choose to display only the voltage signal supplied by a single-mass ion at each time point as a single ion chromatogram (SIC). These chromatograms can be displayed for either SCAN or SIM mode acquisition. The TIC for SIM mode just contains ion current data summed from fewer mass points.

At a given time point, the intensities of the acquired mass spectrum points can be displayed as a mass spectra. While it would be possible to display each fractional mass point to generate a continuous spectrum, it is more common to sum all point intensities around a unit mass and display them as a spectral bar graph at discrete masses, see Figure 1.8. Since a spectrum is distinctive and characteristic for each compound, it can be used as a fingerprint to identify it. The ability to be able to extract a spectrum on the fly allows us to identify compounds as they appear in the chromatogram.

Extracted spectra can then be submitted for library searching by comparison to a spectral database of known compounds. The results are displayed as a series of best matches with a confidence level assigned to each hit. Comparisons are usually begun with the major peaks in a spectrum and move to lesser peaks. Failure to find an acceptable hit in the library databases is becoming increasingly rare as the databases grow. The NIST05 database of environmentally significant compounds contained 190,800 spectra (163,163 compounds), the Wiley Library (8th edition) contains 532,573 spectra, the Stan Pesticide Library listed 340 compounds, and the Pfleger Drug Library contained spectra for 6300 compounds. The NIST and Wiley libraries do not represent that many pure compounds since many are of the same compound run under differing mass spectrometer conditions. To be an exact fit, the mass spectrometers producing the target data must be calibrated and operated exactly in the same way as the sampling system. Multiple target spectra run under different conditions increase the probability of an acceptable hit. Chromatographic artifacts can also change the sample spectra so that it fails to match the target. Trace artifacts can be removed by a software that filters out minor sample mass fragments. Matching criteria can be modified by comparing the target spectra to the sample spectra instead of the other way around.

Failing to find a library match, the sample's fragmentation pattern can be examined and its structure determined by fragmentation analysis. This is a science unto itself and requires a chemist trained in the subject. A very brief introduction to structural interpretation is included in Chapter 12 and an excellent book on the subject by Fred McLafferty is referenced in Appendix E.

The GC/MS data set stored in the computer is a three-dimensional block of data. Each piece of information has three components: signal intensity, mass, and time. Some software offers topological displays of all the information in the array on a two-dimensional surface. Changes in signal intensity with mass and time are displayed as surface changes on the map's hills and valleys. While it is difficult to extract hard numbers from the map, it is very useful for observing trends, detecting impurities on peak shoulders, and for predicting a compound's characteristic fragments in the presence of neighboring peaks. This last point will be very important when setting up target compound identification during quantitation.

8

DATA PROCESSING AND NETWORK INTERFACING

Once we have chromatographic peak data, the data and control computer can be used for compound identification and quantitation. We can automatically determine the amounts of each compound present, positively identify known peaks, and refer unknown peaks to the library database software for identification.

Some software allows control and data processing of multiple GC/MS systems on a single computer. A block diagram for a data/control computer system is illustrated in Figure 8.1. The computer can also be connected to a computer network for data exchange and to avoid transcription errors from re-entering information. When computer systems are retired and replaced, software must exist to retrieve some of the archived data files stored in incompatible data formats.

8.1 PEAK IDENTIFICATION AND INTEGRATION

In automated target compound quantitation, you select and build a table of characteristic mass fragments and their relative signal strengths for each compound to be analyzed. The main identifier mass fragment for a compound

GC/MS: A Practical User's Guide, Second Edition. By Marvin C. McMaster
Copyright © 2008 John Wiley & Sons, Inc.

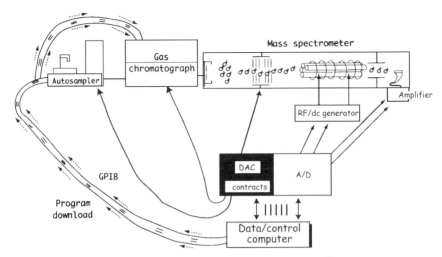

FIGURE 8.1   Data/control computer system diagram.

is called its the target ion; other identifier masses for the same compound
are called qualifiers. When these mass fragments appear in the spectra of
a compound with the correct retention time in the injected sample
chromatogram, we have confirmed the identity of the targeted compound.
You can have further confidence when the relative mass intensity
signals of the target ion and qualifiers agree with the expected target
compound ratios.

Quantitation of the sample components can be done by comparing their
target ion signals intensities to multilevel concentration curves for the target
compounds. Internal standard compounds are added to the analysis mixture to
correct for variations in injection volume and peak retention times on the
column.

In environmental samples, we add surrogate compounds, physically
similar to some of our target compounds, which are used to check for losses
that may occur during extractions. Matrix blanks and matrix spikes are also
included in our analysis deck to determine if materials are added or subtracted
by the sampling matrix as a quality control check on the laboratory's
technique.

In addition to target compounds, internal standards, and surrogates, we
may find other compounds in the analyzed samples. Known compounds are
compounds that we know should be in the original target sample, but we
chose not to analyze. These are marked as such for sample analysis. Other
compounds found in the final analysis sample that do not fit these categories
can be referred to the library for searching as tentatively identified compound

for inclusion in a TIC report. Obviously, these cannot be quantitated, since calibration curves are not available for them.

In preparing a quantitation set, we must first calibrate the mass spectrometer and then run our tuning compound and adjust the lens until we pass a tuning report. At that point, we begin to run our five-level quantitation standards set, each level of which contains target compound standards, internal standards, surrogates, and known compounds. If we have not build our table of target ions and qualifiers for each compound, we take a middle-level quantitation standard run, examine each target peak, surrogate, and internal standard and select a target ion and qualifiers from the largest characteristic fragments. When these are set, we calculate our table of target ions, run our matrix blank and matrix spike sample, and are ready to begin analysis of real samples.

When we are through, the chromatograms and sample reports are examined. If compounds are incorrectly analyzed, because of retention time changes due to column contamination or temperature variations, they can usually be corrected by adjusting the position of the internal standard to which they are compared. In a mixture containing many compounds to be analyzed, multiple internal standards with differing retention times are included. Early running compounds are referenced to an early running internal standard; late runners to late running internal standards. By adjusting the internal standard's retention time, we adjust retention times for all associated peaks. Once all target compounds are analyzed, a series of standard reports can be generated from environmental reporting software.

8.2   MULTI-INSTRUMENT CONTROL

Modern computers are not limited to running a single GC/MS system. The minicomputer-based RTE system software could control two GC/MS systems while acquiring data from both. Quantitation could be handled in a queuing system from data stored on the hard drive. Forms generation was a long, involved batch operation that ground processing and acquisition to a halt. Original specification for the RTE called for it to run four systems simultaneously, but this proved impossible in real-life situations. The latest generation of faster personal computers have made this possible.

To speed operations, many large laboratories began to divide this process across multiple computers. As computer prices dropped and speed and capacity increased, one computer was used to control multiple GC/MS systems. The data files, once acquired, were moved off to a second computer for quantitation that was not slowed by the necessity to time sharing system control and monitoring. A third computer could be used by a quality control monitoring group assigned to make needed chromatographic adjustments. Finally, a fourth computer could be assigned to forms generation.

8.3 NETWORKING CONNECTION

At first, to make all this work, data would be moved from computer to computer on diskette by "sneaker net." Once local area networks (LANs) were set up, the data file could be moved electronically point to point from computer to computer. Eventually, with wide area networks (WANs) in place, files such as reports and data sets could be moved from location to location, to computers of different types, even to facilities in other states.

Care had to be taken that data was not wiped from the original computer until it was certain that the next processing computer had received a copy. Archival storage of the original data set was important when data might have to be defended in court or before a government regulating organization. Large, sequentially stored tape data sets were built to provide this legacy archive. These archives may never be needed, but they must be accessible when the data is required. This has lead to one of the potential problems facing contracting analytical laboratories, which will be discussed next.

8.4 REPLACEMENT CONTROL AND PROCESSING SYSTEMS

Over time, a problem began to arise with the GC/MS system. The mass spectrometer has changed little in the last 15 years, except to become smaller and more compact. Compared to the state-of-the-art computers, the heart of the original processing and control system rapidly became outdated. Hard disks were too small to meet the rising sample demand, the computers were too slow, and the programming was too difficult to use. In addition, the operating system and proprietary data storage system were incompatible with systems from other manufacturers. If a laboratory bought systems from more than one vendor, different systems from the same vendor over time, or acquired other laboratories with different systems, incompatibilities resulted. Only certain operators could run certain machine. If the operators were sick, the machines were sick. If the operator went on maternity leave, the machine was down or someone else had to be trained to run it.

Replacement computers with newer operating systems and better, faster software appeared to be the answer, but replacing a computer was like doing a heart transplant. The new software must be able to control the mass spectrometer, and hopefully, the gas chromatograph and the autosampler. It must be easy to use without requiring massive down time and retraining of the operator. Getting the computer was easy, but finding the software was another matter. Mass spectrometer manufacturers were interested in manufacturing and selling complete systems. Until recently, few had any interest in resurrecting any of their older systems or in upgrading the software that could be used on these systems. Some of the manufacturers were not even still in the mass spectrometry business.

Fortunately, a few third-party companies began to address this question. First, companies appeared that offered data processing on fast modern computers. You still had to run the old control system, but the data handling went much faster. In the last few years, full replacement systems have appeared that replaced the control and data systems on almost any existing GC/MS system. They do full control of the gas chromatography and autosampler programming if the system is capable of remote control. And these system components do not all have to come from the same manufacturer, so you can upgrade other components, such as the GC or the autosampler, as well as the computer. Some original equipment manufacturers have responded by offering upgrade data/control systems, but only for recent GC/MS systems and only for systems where all modules come from their company.

8.5   FILE CONVERSION AND DATA FILE EXCHANGE

Whether on a brand new GC/MS system or on an upgraded system, with the new computer system came a new problem—actually an old problem recognized for the first time—data file incompatibility with the old data files in the archive. Even with a new system or an upgrade from the old manufacturer, data files would probably not be compatible with archived files. For example, the old files might be formatted in Pascal, UNIX, or VMS, while your new files are stored in a Windows or DOS format.

How can the files be changed to the new format? Or do you even want to convert them? What happens when your last tape drive and hard drive that supports the old format bite the dust? Attempts to answer these questions have lead to "elephant burial grounds" of old tape drives and obsolete computer control systems in many facilities trying to maintain access to valuable data archives. They certainly do not want to convert all this old, mostly dead data, simply to have access to a single piece of data.

Again, the third-party companies have come to the rescue. A software company named Chippewa Computing has produced a Pascal-to-DOS conversion software for old Hewlett-Packard systems. A software package called Reflexions for the RTE, will convert its UNIX-based data format to DOS or Windows. There have been rumors of a similar package to convert Finnigan-Mat DEC format to DOS, but this has not been verified. The best way to get information on conversion software is to call technical support at the data system manufacturer.

In cooperation with the American Chemical Society, mass spectrometry organizations have met to define a common database format called ansi-CDF. Each manufacturer would have the responsibility to write translators for each of their data formats into ansi-CDF. You could then translate data sets from one proprietary format to another by going through this "bridge format."

A similar CDF format was proposed to all chromatography companies facing similar problem, but was largely ignored for years until the growth in importance of laboratory information systems (LIMS). LIMS require that you be able to import data from a variety of wet chemical and chromatography systems in order to generate standard laboratory reports on analyzed compounds. This drive toward common results reporting has lead to a strong movement toward common metafile and data formats. The key here is to follow the money. Companies have no financial incentive to cooperate and common data files may open the door for a competitor into a laboratory they already control.

Data file exchange between data systems from different manufacturers continues to be a problem in laboratories with older GC/MS systems from different manufacturers and often with system of different ages from the same manufacturer. Environmental reporting requirements of the US Environmental Protection Agency has forced a certain amount of standardization in this field. They set up reporting requirements for their contract laboratory program (CLP). A laboratory working within this program must submit these forms in paper and as disk deliverables. Equipment manufacturers and third-party software companies wrote software designed to take the output from various GC/MS systems and processes it into CLP-type reports.

Because of the availability of this software, state and local regulatory agencies and commercial companies with internal environmental testing programs adapted the federal requirements to their own need. Even though there were only 10 contract laboratories in 1995, currently these report types are used through out the United States and overseas. Often local company modifications have loosened some of the requirements of the CLP program for use in their own laboratories. CLP reporting requirement have also undergone modification and fine tuning about ever other year. But CLP-type reports remain the *de facto* standard for the industry and the same type of reporting can be produced from almost any GC/MS system.

8.6   DATA RE-ENTRY AND TRANSCRIPTION ERRORS

We will assume that we can process the data into chromatograms and spectra and generate the required reports from the raw data after verifying the performance of the system. But, getting these data into a form for reporting to customers has proved a problem for commercial laboratories. The laboratory reports have to be combined with records and data on the sample and put in a form that the customer can understand and use. The usual way that this is done is that the data from various reports is abstracted and retyped into a final report format. The problem that often occurs is that data get reentered incorrectly. Transcription errors can be very expensive, both to the customer and eventually to the commercial laboratory both in terms of time and money. Results have to be re-examined and corrected reports sent out. If this problem occurs too often, it is very easy to lose important customers in this very competitive industry.

To avoid, or at least minimize, the problem, many laboratories are moving to computer networked based LIMS. A LIMS system is designed to prompt and check data type for manual input of sample information, such as source, ID numbers, and testing requirements. It will then pull information automatically from report files on a variety of laboratories computers to generate a final customer report without further operator intervention. Transcription errors are avoided, and once the system is running, report inspection is minimized. Setting up the system to extract exactly the needed report file components is much harder and time consuming than actually running the software.

A LIMS system is only as good as its input data. It can detect missing information and flag it as an error, but incorrect data still generate incorrect final output. It is still necessary to have visual review and quality control of the data report information to assure the quality of the output going to the customer.

9

SYSTEM MAINTENANCE AND TROUBLESHOOTING

9.1 GAS CHROMATOGRAPH MAINTENANCE

The most critical part of the gas chromatograph is the injection port. Approximately 90% of all chromatography problems can be traced to this component. This is where vaporization of the sample takes place. Maintaining inertness and a leak-free environment are top priorities in a laboratory where difficult sample matrices are analyzed (such as base-neutral aromatics, BNAs). The following tasks should be performed daily.

9.1.1 Injector Port and Liner

Replace the injection port liner with a fresh liner daily. Liners should be silanized to remove active sites in the glass. Also a small "wisp" of silanized glass wool should be inserted into the liner about half way. The glass wool increases the available surface area in the liner, which helps to promote vaporization. It is also effective in filtering nonvolatile residues and preventing them from making their way to the head of the column.

GC/MS: A Practical User's Guide, Second Edition. By Marvin C. McMaster
Copyright © 2008 John Wiley & Sons, Inc.

9.1.2 Septum Replacement

Replace the septum every 2–3 days. Septa are good for approximately 100 injections. If you fail to change your septum regularly, it will begin to leak, causing retention time shifts and column bleed.

On Hewlett-Packard gas chromatographs, replace the rubber O-ring around the liner and the injection port disk inside the nut at the base of the injection port.

9.1.3 Syringe Cleaning

Manual injection syringes should be rinsed at least twice with solvent before filling. Once a day rinse with an intermediate-polarity solvent such as tetrahydrofuran (THF) or methylene chloride to insure that precipitated nonpolar compounds are not coating the barrel. If particulate from evaporation plug or block the syringe, ream it out with the fine wire supplied in the syringe box.

An autosampler syringe should be cleaned daily. Remove the syringe from its holder and rinse the plunger with methylene chloride. Also try to introduce some methylene chloride into the shaft of the syringe.

For column maintenance, I recommend using a guard column when ever possible. A guard column will collect nonvolatile residues that would otherwise accumulate at the head of the column. These residues can interfere with chromatography. If you do not use a guard column, then I recommend trimming 6–12 cm off of the head of the column daily. Also, be sure to use a new ferrule when you reinsert the column into the injection port. The column should be inserted 5 mm into the injection port.

Be sure not to exceed the column manufacturers maximum allowable temperature as this will cause column bleed and or coating collapse, shortening the column's life. Also, guard against leaks, as oxygen will strip the stationary phase of the nonbonded column.

All fittings should be tightened one-half turn past finger tight. If you go much farther than this, the ferrule will fail and begin to leak.

9.1.4 Carrier Gas Selection and Purification

Helium is the most commonly used carrier gas, although it is expensive. It is inert and has low viscosity for good chromatographic separations. Hydrogen is cheap and has low viscosity, but it is also quite explosive. Health and safety coordinators of most laboratories will not approve its use. Nitrogen is inexpensive, inert, and nonflammable, but it has higher viscosity and thus is not an efficient carrier gas.

No matter what carrier gas you use, you will need to use an oxygen trap. The trap not only prevents trace amounts of oxygen present in the carrier gas from reaching the column but also collects oxygen from leaks that may be present in the fittings. The trap should be changed every 6 months at least, depending on the type of trap.

9.2   MASS SPECTROMETER MAINTENANCE

The two major problems in day-to-day mass spectrometer operation beyond reaching and holding a tune are air leaks and source burning. Most of your attention will be devoted to the instrument's source. Instrument manufacturers all have taken their own approach to source design, the objective basically being to create ions and to tune the source lens so that desired relative peak mass ratios and resolutions can be achieved. But no matter what instrument you have, you will have to clean the source from time to time. How often depends on rate of use, the nature of the samples you are analyzing, and the frequency specified by your protocol.

9.2.1   Problem Diagnostics

If you are having trouble reaching full vacuum but the filament ignites, check for air leads by scanning from m/z from 0 to 50 amu. Look for the water (18), nitrogen (28), and oxygen (32) peaks. If they are present, there probably is a leak around the column-to-source seal. Shut down and check your fitting and ferrule. If the fitting is snug, the ferrule is probably scored and needs replacing. Also check the seal around the calibration gas valve. Occasionally, there are air leaks here. Often briefly opening the valve and then reseating it can eliminate this source of leaks.

An excellent method of determining when you need to do source maintenance is to monitor the calibration gas 502 fragment height from autotuning. Measure the height after tuning a new or freshly cleaned instrument. Set an acceptable threshold, say 10% of the clean level. Once the 502 value drops below this minimum standard, it is time to clean. The 502-fragment is chosen because heavier fragments are much more easily effected by dirty or corroded source surfaces.

Once your have decided to clean the source, vent the analyzer, power down, and cool the gas chromatograph. Next, remove the column and the interface from the mass spectrometer. The analyzer source assembly is removed from the analyzer and disassembled. The ion and filament contract surfaces must be cleaned and dried. The ion source assembly is reassembled, inserted back into the analyzer, and leads are electrically reattached. The interface and column are reconnected and the analyzer re-evacuated and all temperature zones reheated. Finally, an autotune is run to establish that the 502 fragment height is back over the performance threshold.

The venting and power-down sequence will vary from instrument to instrument. It is important to follow exactly the procedure indicated in the instrument manual. Turbomolecular pumps are designed to be turned off at speed and evacuated through the rough pump. Oil diffusion pumps must be cooled to less than 100 °C before venting or they will backstream oil into the analyzer, contaminating the quadrupole surfaces. Once cooled, carefully feel the temperature of the pump exhaust, it is a pretty reliable guide that the diffusion pump is cool enough to turn off.

Once the GC oven is cooled and the column and interface removed, it is important to protect the MS interface insertion surface. Remove it from the analyzer and protect it by wrapping it in aluminum foil until it is time to insert it back into the analyzer section.

9.2.2   Source Cleaning

To clean the source, disassemble it in a clean area where you have plenty of room to work. Take special care not to lose small parts. Remove the control interface cables and the electrical connections to the filaments, the repeller, and the various focus lenses. Unfasten and remove retaining screws that hold the filaments and the repeller to the source body. On a HP 5972 mass spectrometer, the whole source assembly up to the entrance lens can be removed as a single piece for disassembly and cleaning.

Be very careful not to use abrasives on Vespal surfaces or allow solvent to get under these surfaces. These surfaces tear up easily and solvent will cause them to swell, making reassembly difficult. It is usually easy to recognize these colored, plastic-looking surfaces.

If you are in doubt, do not clean them. Burn and char are usually pretty obvious. Both type of char result from high temperature oxidation of nitrogen-containing organic compounds. The largest molecules in a separation come out of the gas chromatograph at the hottest end of the temperature ramp. They then hit an evacuated volume, where they are bombarded with 70-eV electrons.

The pieces that need cleaning are those in contact with the ion stream: the repeller face, the ion source inner body, both sides of the draw-out plate and its pinhole entrance, the focus lenses, and entrance lens contact surfaces. The ion source body shows burn next to where they contact the filaments. Filaments need to be removed and the holes leading into the source body need to be cleaned by reaming with a fine drill.

Cleaning source surfaces is an art and a source of controversy. A variety of abrasive, chemical, sonic, and electroplating techniques have been described in the literature. Abrasive techniques using fine powders are effective, require minimum equipment, and are reasonably safe for the source surfaces. Hewlett-Packard recommends cleaning surfaces with aluminum oxide powder and methanol paste for use with Q-tips. It also supplies aluminum oxide paper that can be used in cleaning inner surfaces and drill bits and a holder for reaming out pinhole entrances. Some laboratories consider aluminum oxide too harsh and use the less abrasive jeweler's rouge for cleaning. Also, there are laboratories that use a rouge paste in a tube that is intended for motorcycle detailing. Instrument service and repair facilities use high pressure sand and water blasting techniques to clean mailed-in sources.

Once the source elements are disassembled, the flat lenses can be cleaned lightly with a jeweler's paste and a soft pad on a Dremel tool; under no circumstances should abrasive stones or rubber wheels be used. These have the potential of scarring the surface, which can alter the electrostatic field of the lens. Once scarring occurs, element may not perform as well as originally intended. Parts that are too small for the Dremel tool should all be placed in a small beaker. The larger flat pieces should also be placed in a larger beaker. Fill the beakers with water and add a few drops of Alquinox (an industrial soap). Place the beakers in a sonicator and sonicate for approximately 1 h. After 1 h, remove the beakers and very carefully pour out the water (be especially careful not to lose any parts). Then refill the beakers with methanol to remove the residual water. Sonicate again for about 5 min. Remove the parts from the beakers and dispose of the methanol in an appropriate waste solvent container.

Other rinsing procedure calls for sonication in a series of solvents. After wiping all surfaces with Q-tips to remove as much abrasive as possible, it is recommended to sonicate for 5 min twice each in a chlorinated solvents such as methylene chloride, then acetone, and finally methanol. Then air dry, place all parts in a beaker, and dry in an oven at $100\,°C$ for 15 min.

Next, lay out the parts and inspect their cleanliness. At this point, you should wear thin cotton gloves to prevent parts from finger grease. If the source has small ceramic collars or spacers, inspect them for cracks or chips. It would be a good idea to replace these as necessary. Maintain a good spare-parts inventory to cover this. Reassembly is the reverse of disassembly. Reassemble the source body including the insulators around the ion focus and entrance lenses. Insert the source body back into the analyzer body after reattaching the repeller and the filaments. Make sure that the ceramic collar between the source and the quadrupole does not bind; it must turn freely. Connect the repeller and filament leads as well as leads to the focus lenses.

When your source is assembled, you should conduct an electrical continuity check to insure that first, there is appropriate continuity and, second, that no shorts exists. You are then ready to reinsert your analyzer into the mass spectrometer body.

Once the interface is reinserted and electrically connected to the gas chromatography, reinsert the chromatography column, ensuring that proper insertion depth is achieved. How this is done depends on the source design. Generally, you need to slide the column compression fitting and Vespal ferrule onto the column, insert it into the interface until it bottoms out, retract the column about 2 cm, then tighten the compression fitting one-half turn past finger tight.

Replace the housing on the vacuum containment vessel. Turn on the rough pump and begin evacuation. Set your interface heater to its operating temperature, usually around 280 °C. Set your GC temperature zones to their startup values. When the gauge shows that you are at 10^{-4} Torr, turn on the turbo or the oil diffusion pump. If it is a diffusion pump, turn on the pump heater and bring it to the desired temperature. Pump until the normal high vacuum is reached, which will take about 4 h on a HP 5972 with a diffusion pump. This will vary for other systems.

If there is a problem establishing the initial rough vacuum, push down on the lid on the containment vessel. This is all that is necessary in most cases. If you still have a problem, stop the vacuum, inspect, and clean the gasket around the inside of the lid. If necessary, replace the gasket.

Once full vacuum is reached and you have checked for air leaks by scanning below 50 amu for water, nitrogen, and oxygen, check the effectiveness of your cleaning procedure. Rerun you autotune procedure and check the 502 peak height. It should now be somewhere between the instrument's best measurement and your system's minimum performance level.

9.2.3   Cleaning Quadrupole Rods

Another potential problem that can occur in the quadrupole that will effect its operation is the accumulation of organics on the quadrupole rods. Ions that do not survive travel through the quadrupole collide with the rods, pick up an electron, and become electrically neutral molecules. If they are small volatile compounds, they are swept off the rod by the vacuum system and end their life as oil contaminats in the vacuum pumps. However, there is a slow accumulation of larger, nonvolatile organics on the quadrupoles. Periodically, these must be washed off since they will distort the electromagnetic field and eventually shut down the analyzer. In order to remove the rods for cleaning, the system must be vented and the source removed as before. The ceramic collar between the source and the quadrupole is removed, electrical connections to the rods are disconnected, and the rod package is removed.

Note. The rods on most systems are held in exact hyperbolic alignment by two ceramic collars that *must not be* removed. Nothing will shut a mass spectrometer down faster than messing up rod alignments. The minimum that must be done to repair this problem is to ship the rod package in for repair and realignment. This is time consuming and expensive and not always successful.

Rods are cleaned by immerging the complete quadrupole package of four rods in their ceramic collars in a graduated cylinder and flooding it with solvent. You must be very careful not to chip a rod while placing them in the cylinder. Older quadrupole systems have very large rod packages, and they are usually cleaned by removing the rod package and wiping them with large lintless paper towels. Usually they are washed first with a nonpolar solvent such as hexane, then with methylene chloride, and finally with dry acetone. Modern quadrupoles can be air dried and then evacuated in a desiccator if the rod package is small enough to fit. Oven drying at 100 °C has been rumored to cause rod distortion because of differential expansion of the collars and rods, and probably should be avoided. Air dry large rod packages as well as possible; then finish the drying process as the rods are evacuated in the mass spectrometer vacuum housing. It will not help your roughing pump ail, but it will dry the rods.

9.2.4 Ion Detector Replacement

Ion detector horns have a finite lifespan and should be replaced when noise begins to increase. Run the repeller to its maximum value, and then look at the electron multiplier voltage necessary to get a 502 fragment for calibration gas above the benchmark value. When the EM voltage gets above 3500 V, it is time to consider replacing the detector.

9.2.5 Pump Maintenance and Oil Change

The roughing pumps are the only pumps you will be expected to service. Oil in these pumps should be changed every 6 months. Change the oil when it becomes brown and cloudy on inspecting the viewing port on the side.

Hoses coming from the roughing pumps should be periodically checked for cracks. Hose thickness and diameter are critical for proper performance of your vacuum system. Avoid the temptation to substitute other size tubing when you need to do emergency replacements.

Trained technicians from the manufacturer should service oil diffusion and turbomechanical pumps. Most manufacturers offer some type of trade-in program for turbo pumps and this should definitely be part of the purchase agreement when buying a system. When these systems go down, rebuilding them is a major undertaking. They operate at very high speeds with little tolerance for error. When they are down, they are down.

9.3   SYSTEM ELECTRICAL GROUNDING

This is a problem that laboratories should never face. Grounding problem should have been worked out by the manufacturer before and during system installation and should not occur unless the system is altered. However, systems do wear out, they are moved, and changes are made. Grounding problems can occur when replacing the controlling software and interfaces and moving to modern computers.

For a while I demonstrated and sold replacement systems for a number of types of mass spectrometers. In a few cases, I saw problems with calibration gas peaks failing to stabilize. They jumped from side to side of the expected position. The problems only disappeared when the mass spectrometer chassis and the controlling interface were connected with a grounding strap and all electrical and computer systems were joined through a common surge protector. The problem seemed more common during winter months when laboratories are particularly dry. I suspect that static electric discharges may be involved, since I have seen similar problem with other types of analytical systems during the dry winter months in St. Louis laboratories.

A similar problem was seen and cured on a system in Minnesota that was thought to have been subjected to a nearby lightening strike. The system had its original interface and computer system. The calibration problem was very similar to that seen after connecting a demonstration systems and was cured by grounding the interface to the mass spectrometer.

When this problem occured it was very frustrating. You could almost calibrate the system, but it would never hold a tune sufficient for environmental analysis.

PART III

SPECIFIC APPLICATIONS OF GC/MS

10

GC/MS IN THE ENVIRONMENTAL LABORATORY

One of the major uses of GC/MS systems is in environmental testing laboratories. These can be commercial testing laboratories that do environmental testing for the public on a fee-for-test basis. They can be industrial in-house laboratories that do standard industrial testing as well as additional specific testing for their company's products and by-products. They can also be laboratories participating in the EPA's Contract Laboratory Program (CLP) that do testing for the EPA regional laboratories. Most of these laboratories will be working with methods developed, standardized, and periodically updated by the US EPA for their CLP laboratories. Laboratories not involved with the CLP program may modify the standard programs for their own use. Many will add compound of interest to the lists of volatile organic analysis (VOA) and semivolatile organic analysis (semi-VOA) compounds. Some laboratories do not use the stringent tuning requirements of the EPA methodology. For example, the standard VOA requires passing bromofluro-benzene (BFB) tuning within the three scans closest to the BFB peak maximum. This is done to avoid contaminants that most often occur at the chromatographic peak front or on the tailing back side. Other laboratories only require that you pass BFB tune with a scan anywhere within the BFB chromatographic peak. This probably seems to be a trivial change, but it can make a big difference in the time necessary to tune older mass spectrometers.

GC/MS: A Practical User's Guide, Second Edition. By Marvin C. McMaster
Copyright © 2008 John Wiley & Sons, Inc.

When you first look at environmental methods, you are met with a bewildering list of numbered methods. Almost all are based on one of the two procedures, with variations in the basic methods to allow for differences in the sample matrix. VOA is used to analyze low molecular weight compounds that can be purged out of an aqueous solution with an inert gas and captured on a solid packing. Semi-VOA is used to quantitate larger, less volatile organic compounds that must be extracted from the matrix before they can be injected, separated in the gas chromatograph, and analyzed in the mass spectrometer. VOA analyses are carried out on drinking water, wastewater, hazardous waste, and air monitoring samples. Semi-VOA analyses are made on drinking water, wastewater, and hazardous materials.

Other available GC/MS sample analysis methods used in some of these laboratories include dioxan/furans and pesticide/PCB confirmation. These represent a much smaller volume of work for the environmental laboratories. The analysis of dioxan/furans is also an extracted method and could be considered a semi-VOA analysis, but it is a very specific method with hazardous sample handling and is not routinely done by all environmental laboratories. Pesticides/furans have an EPA-approved GC-only method; some laboratories used the GC/MS confirmation test to provide a definitive proof that coeluting compounds are not confusing the GC-only method.

We will take a broad look at VOA (method 624) and semi-VOA (method 625) analysis of wastewater as representative of these techniques. For specific details of environmental analysis procedure, you should always consult the latest US EPA versions. These are available in print form directly from the EPA and in diskette and CD versions from various suppliers for use with your analysis computer.

10.1   VOLATILE ORGANIC ANALYSIS: EPA METHOD 624

The EPA method 624 will be discussed as a typical VOA. It is an analysis of the volatile organic components of wastewater using a purge-and-trap apparatus to introduce sample into the GC/MS system. The compounds identified and quantified are halogenated hydrocarbons and aromatic hydrocarbons with boiling points below $200\,^\circ$C and molecular weights below 300 amu. Sample must be pulled, stored in amber glass bottles at $4\,^\circ$C, and analyzed within 14 days.

Methods 524 for drinking water, 8240 for hazardous waste, and CLP 2/88 and 3/90 VOA methods are all similar techniques with changes made for different sample matrix or reporting requirements.

In summary, the mass spectrometer running in EI mode must be calibrated along its mass axis using PFTBA calibration gas and tuned for BFB analysis using an autotune method or by hand, if no autotune is available. The tune is checked by injection of BFB solution through the gas chromatograph to see if the correct mass fragment ratios are produced (Table 10.1).

Once the GC/MS passes the BFB tune report, a standards calibration run must be made. If this is the first tune of the day, five concentration levels of the mixture of 624 VOA standards, surrogates, and internal standards are run and response factors calculated. If this is not the first tune of the day, a single-level continuing calibration (CC) standard run must be made and compared to the last five-point standards run and should fall within the acceptable variation range. If it fails to calibrate, then the five-point standards must be rerun.

Both surrogates and internal standards are spiked into the sample to be purged. Surrogates are compound with chemical structures similar to the standard compounds, usually deuterated or fluorinated compounds, that are not found in nature and are added as a check on purging recovery. Internal standards are compounds with similar chromatographic behavior not affected by the method or matrix. Response factor for each compound are calculated for each target compound relative to the internal standards from the calibration standard runs.

When acceptable standards are achieved, then blanks and samples can be run. A 5-ml analysis sample or blank is placed in the purge-and-trap apparatus illustrated in Figure 2.2 and sparged with 40 ml/min of helium for 11 min at ambient temperature into a baked sorbent trap predried with purge gas.

The trap is packed from the inlet end with equal amounts of Tenax (a porous, cross-linked resin based on 2,6-diphenylene oxide), then silica gel, and then charcoal. Organics are trapped in the Tenax, water in the silica gel.

TABLE 10.1   BFB *m/z* Abundance Criteria

Mass	*m/z* abundance criteria
50	15–40% of mass 95
76	30–60% of mass 95
95	Base peak, 100% relative abundance
96	5–9% of mass 95
173	<2% of mass 174
174	50% of mass 95
175	5–9% of mass 174
176	95% but <101% of mass 174
177	5–9% of mass 176

The 7-1-87 EPA protocol added a 1-cm plug of methyl silicone (OV-1) before the Tenax to protect and refresh the column, but omitted the charcoal plug at the end. Unretained purge gas is vented to atmosphere through a vent valve at the rear of the trap.

After trapping is complete, the purge gas diverter valve is turned and helium is introduced from the exit end of the trap. The trap is rapidly heated to 180 °C and back flushed with purge gas to desorb the organic components into the gas chromatographic column. The water trapped in the silica gel explodes into steam, helping to desorb the organics from the Tenax, but must be diverted away from the GC column with a valve into a dryer.

The GC column specified for the method is 1% SP-1000 on Carbopack-B packed in a 6 ft × 0.1 in. column. However, in recognition of advances in chromatography techniques, capillary and bonded-phase columns can be used. In selecting these methods, the analyst must adjust the analysis conditions to bring the method into compliance with the expected standards' separation.

Due to the chemical stability of bonded-phase capillary column, most analysts are moving to these column to avoid column bleed into the mass spectrometer.

A GC oven temperature ramp program is run to elute chromatographic peaks into the mass spectrometer where they are scanned for identification and quantitation. The injector temperature is set at 200 °C and helium carrier gas flow is adjusted to 10 ml/min. The initial GC oven temperature of 35 °C is held for 6 min: then a 10 °C/min ramp is run to 210 °C, where it is held for 5 min. The sample passes through the interface heated to 200 °C into the mass spectrometer operated at 70 eV, which is scanned from 45 to 300 amu.

Table 10.2 shows VOA standard's retention times on a DB624 column 75 m long with a 3.0 mm film thickness. Also shown are the primary ions, secondary ions, and experimentally determined minimum detection levels (MDLs) calculated from seven replicates.

While the run is being made, the purge residue is flushed out of the purge tube with purge gas, rinsed two times with reagent water, and blown dry. The trap is baked out at 180 °C with fore flow of purge gas to vent, preparing it for the next sample.

The oven temperature must be returned rapidly to the injection temperature and equilibrated for the next injection. Automated purge-and-trap apparatus is available from a number of companies so that a series of standards and/or samples can be prepared for sequential analysis.

The quantitation software database is prepared ahead of the time with data from a middle-range calibration standard. Response factors and retention times are calculated from standard runs, and primary and secondary target

TABLE 10.2   VOA Target Compounds

Compound	Retention time, min	Primary ion	Secondary ion(s)	MDL, μg/l
Chloromethane	7.84	50	52	2.1
Vinyl chloride	8.33	62	64	1.6
Chloroethane	9.99	64	66	1.3
1,1-Dichloroethene	12.59	61	96, 98	3.0
Methylene chloride	13.88	84	49, 86	2.1
1,2-*trans*-Dichloroethane	14.58	61	96, 98	1.6
1,1-Dichloroethane	15.53	63	65, 83	1.6
Chloroform	17.38	83	85	1.6
1,1,1-Trichloroethane	17.83	97	99	1.4
Carbon tetrachloride	18.18	117	119, 121	1.5
Benzene	18.52	78	77	1.6
1,2-Dichloroethane	18.53	62	64, 98	3.0
Trichloroethane	19.69	130	95, 132	2.1
1,2-Dichloropropane	20.11	63	65	1.8
Toluene	21.99	91	92	1.9
1,3-Dichloroprpane	21.34	75	110, 112	2.1
1,1,2-Trichloroethane	22.60	97	83, 85	2.5
Tetrachloroethane	23.00	166	129, 164	1.9
Chlorobenzene	24.52	112	77, 114	1.6
Ethylbenzene	24.64	91	106	2.2

ions for each target compound are entered (Table 10.2). Data from sample and blank runs are processed with this information to determine the identity and amounts of each target compound present. Known compounds, which will not be quantitated, surrogates, and internal standards are all marked in the quantitation software. Amounts of target compounds in matrix blanks, reagent blanks, QC check samples, and matrix spike samples are calculated for various quality control reports. Unknown compound, which are not target compounds, surrogates, internal standards, or know compounds, are identified, referred to library searching, and reported as tentative identified compounds in a TIC report.

Basic total-ion chromatograms need to be inspected by quality control before final reports are made. Compound retention times are required to be within a 30-s window, target masses must maximize within one scan of each other, and relative fragment mass peak heights must fall within 20% of those for a reference spectrum. When calibration standards are run, retention times can be adjusted relative to the retention of internal standards to correct for variations in the column-separating characteristics. If sample or blank retention times fall outside these windows, the column must be modified or replaced and standards reruns.

Quality control is very important. An initial method blank and standard spike in reagent water are required to demonstrate the laboratory's cleanliness and ability to run within parameters. Samples must be spiked with standards and reanalyzed for 5% of the total samples analyzed as a performance check. Check standards available from the EPA must be run periodically to demonstrate the laboratory's capability. A schematic of method 624 VOA analysis is shown in Figure 10.1.

10.2   SEMIVOLATILE ORGANIC ANALYSIS: EPA METHOD 625

Semi-VOA analysis is done by extracting base-neutral compounds from the grab sample with methylene chloride after pH adjustment to >11 with sodium hydroxide, followed by acidification with sulfuric acid to pH <2.0 and re-extraction with methylene chloride. Grab samples must be stored in glass containers at 4 °C and extracted within 7 days and completely analyzed within 40 days.

Method 625 is for wastewater samples, 525 for drinking water, and 8250/8270 for hazardous waste. CLP methods for these were modified in 2/88 and 3/90 with other method changes expected periodically. Contact your regional US EPA for current methods. Changes are generally to allow use of more stable columns and more sensitive detectors for trace impurities detection.

Base-neutral compounds analyzed by method 625 are halogenated aromatics, nitro aromatics, polynuclear aromatics, aromatic ethers, pesticides, and PCBs. The possible presence of dioxins can be analyzed as part of this method, but they must be conclusively determined using EPA method 613. Compounds analyzed as acid extractables are substituted phenols. All compounds must have fragment masses below 450 amu.

Sample preparation is done by adding surrogate compounds to the analysis sample and blanks before extraction. The pH is adjusted to >11 with 10N sodium hydroxide and the sample extracted twice with methylene chloride to yield a base-neutral fraction. The aqueous fraction is then acidified to pH <2 with 50% sulfuric acid and extracted twice more with methylene chloride to yield an acids fraction. The base-neutral and acids fractions are analyzed separately. Each is dried over sodium sulfate and concentrated using Kuderna–Danish evaporators to remove solvent. Internal standards are

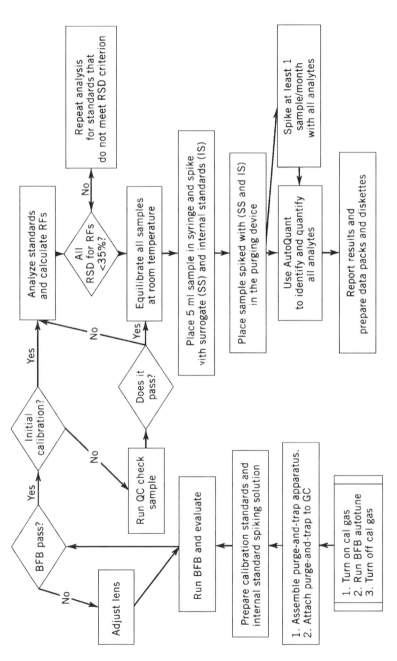

FIGURE 10.1 Method 624 VOA schematic.

TABLE 10.3   DFTPP _m/z_ Abundance Criteria

Mass	_m/z_ abundance criteria
51	30–60% of mass 198
68	<2% of mass 69
70	<2% of mass 69
127	40–60% of mass 198
197	<1% of mass 198
198	Base peak, 100% relative abundance
199	5–9% of mass 198
275	10–30% of mass 198
365	>1% of mass 198
441	Present, but less than mass 443
442	>40% of mass 198
443	17–23% of mass 442

added to the concentrates, which are made up to injection volume and placed in autosampler vials.

After the mass spectrometer is tuned using a decafluorophenylphosphine (DFTPP) autotune, it is set up for scanning from 35–450 amu and a sample of DFTPP solution is injected into the gas chromatograph. A DFTPP tune check is run using the abundances in Table 10.3.

If the tune check passes all criteria, base-neutral calibration standards or a continuing calibration standard is run. If the tune check fails, the lens are readjusted and the DFTTP injection repeated until the DFTTP check passes. Once passed, the GC/MS system is certified to run samples for 12 h after which it must be recertified with DFTTP.

The column used for base-neutral extractables is 1.8 m × 2 mm internal diameter packed with 3% SP-2250 on Supelcoport support. The method requires that a sample of benzidine must be injected and a tailing factor calculated. The column must be replaced if the tailing factor criteria cannot be achieved. One of the advantages of the capillary column is that removing a small portion on the inlet end of the column can aid in passing these tailing criteria and extend the lifetime of the column. If this is the first calibration, standards are run and response factors are calculated for all target compounds against internal standards.

The acidic extractable column is a 1.8 m × 2 mm column packed with 1% SP-1240A on Supelcoport packing. It must be evaluated by injection with pentachlorophenol and checked for tailing. Once it passes, standards are run and response factors calculated or a continuing calibration standard must be passed.

TABLE 10.4   Semi-VOA (Base-Neutral) Target Compounds

Compound	Retention time, min	Primary ion	Secondary ion(s)	MDL, μg/l
1,3-Dichlorobenzene	7.4	146	148, 113	1.9
1,4-Dichlorobenzene	7.8	146	148, 113	4.4
Hexachloroethane	8.4	117	201, 199	1.6
bis(2-Chloroethyl)ether	8.4	93	63, 95	5.7
1,2-Dichlorobenzene	8.4	146	148, 113	1.9
bis(2-Chloroisoproply)ether	9.3	45	77, 79	5.7
N-Nitrosodi-n-proplyamine	—	130	42, 101	—
Nitrobenzene	11.1	77	123, 65	1.9
Hexachlorobutadiene	11.4	225	223, 227	0.9
1,2,4-Trichlorobenzene	11.6	180	182, 145	1.9
Isophorone	11.9	82	95, 138	2.2
Naphthalene	12.1	128	129, 127	1.6
bis(2-Chloroethoxy)methane	12.2	93	95, 123	5.3
Hexachlorocyclopentadiene	13.9	237	235, 272	—
2-Chloronaphthalene	15.9	162	164, 127	1.9
Acenaphthylene	17.4	152	151, 153	3.5
Acenaphthene	17.8	154	153, 152	1.9
Dimethyl phthalate	18.3	163	194, 164	1.6
2,6-Dinitrotoluene	18.3	165	89, 121	1.9
Fluorene	19.5	166	165, 167	1.9
4-Chlorophenyl phenyl ether	19.5	204	206, 141	4.2
2,4-Dinitrotoluene	19.8	165	63, 182	5.7
Diethyl phthalate	20.1	149	177, 150	1.9
N-Nitrosodiphenylamine	20.5	169	168, 167	1.9
Hexachlorobenzene	21.0	284	142, 249	1.9
β-BHC	21.1	183	181, 109	—
4-Bromophenyl phenyl ether	21.2	248	250, 141	1.9
δ-BHC	22.4	183	181, 109	—
Phenanthrene	22.8	178	179, 176	5.4
Anthracene	22.8	178	179, 176	1.9
β-BHC	23.4	181	183, 109	4.2
Heptachlor	23.4	100	272, 274	1.9
δ-BHC	23.7	183	109, 181	3.1
Aldrin	24.0	66	263, 220	1.9
Dibutyl phthalate	24.7	149	150, 104	2.5
Heptachlor epoxide	25.6	353	355, 351	2.2
Endosulfan I	26.4	237	339, 341	—
Fluoranthene	26.5	202	101, 100	2.2
Dieldrin	27.2	79	263, 279	2.5
4,4'-DDE	27.2	246	248, 176	5.6

(*continued*)

TABLE 10.4 *(Continued)*

Compound	Retention time, min	Primary ion	Secondary ion(s)	MDL, µg/l
Pyrene	27.3	202	101, 100	1.9
Endrin	27.9	81	263, 82	—
Endosulfan II	28.6	237	339, 341	—
4,4'-DDD	28.6	235	237, 165	2.8
Benzidine	28.8	184	92, 185	44
4,4'-DDT	29.3	235	237, 165	4.7
Endosulfan sulfate	29.8	272	387, 422	5.6
Endrin aldehyde	—	67	345, 250	—
Buthyl benzyl phthalate	29.9	149	91, 206	2.5
(bis)2-Ethylhexyl) phthalate	30.6	149	167, 279	2.5
Chrysene	31.5	228	226, 229	2.5
Benzo(*a*)anthracene	31.5	228	229, 226	7.8
3,3'-Dichlorobenzidine	32.2	252	254, 126	16.5
Di-*n*-octyl phthalate	32.5	149	—, —	2.5
Benzo(*b*)fluoranthene	34.9	252	253, 125	4.8
Benso(*k*)fluoranthene	34.9	252	253, 125	2.5
Benzo(*a*)pyrene	36.4	252	253, 125	2.5
Indeno(1,2,3-*cd*)pyrene	42.7	276	138, 277	3.7
Dibenzo(*a,h*)anthracene	43.2	278	139, 279	2.5
Benzo(*ghi*)perylene	45.1	276	138, 277	4.1
N-Nitrosodimethylamine	—	42	74, 44	—
Chlordane[a]	19–30	373	375, 377	—
Toxaphene[a]	25–34	159	231, 233	—
PCB 1016[a]	18–30	224	260, 294	—
PCB 1221[a]	15–30	190	224, 260	30
PCB 1232[a]	15–32	190	224, 260	—
PCB 1242[a]	15–32	224	260, 294	—
PCB 1248[a]	12–34	294	330, 262	—
PCB 1254[a]	22–34	294	330, 362	36
PCB 1260[a]	23–32	330	362, 394	—

[a]These compound are mixtures of various isomers.

Now that standards are set, blanks and sample can be run on their respective columns. Base-neutrals are injected with helium carrier gas at 40 ml/min flow rate. The column equilibrated at 50 °C is held for 4 min and then a 8 °C/min ramp is run to a final temperature of 270 °C and held at the final temperature for 30 min. Table 10.4 shows a list of the base-neutral extractables, their retention times, MDLs, and primary and secondary fragment ions.

TABLE 10.5   Semi-VOA (Acid-Extractables) Target Compounds

Compound	Retention time, min	Primary ion	Secondary ion(s)	MDL, µg/l
2-Chlorophenol	5.9	128	64, 130	3.3
2-Nitrophenol	6.5	139	65, 109	3.6
Phenol	8.0	94	65, 66	1.5
2,4-Dimethylphenol	9.4	122	107, 121	2.7
2,4-Dichlorophenol	9.8	162	164, 98	2.7
2,4,6-Trichlorophenol	11.8	196	198, 200	2.7
4-Chloro-3-methylphenol	13.2	142	107, 144	3.0
2,4-Dinitrophenol	15.9	184	63, 154	42
2-Methyl-4, 6-dinitrophenol	16.2	198	182, 77	24
Pentachlorophenol	17.5	266	264, 268	3.6
4-Nitrophenol	20.3	65	139, 109	2.4

The acid-extractable compounds are injected with helium carrier gas at 30 ml/min flow rate. The column is held isothermal at 50 °C for 4 min and then a 8 °C/min ramp is run to a final temperature of 200 °C and held at the final temperature for 5 min. Table 10.5 shows a list of the acid-extractables, their retention times, MDLs, and primary and secondary fragment ions.

As with VOA sample, total-ion chromatograms are examined to insure retention times, ion fragment masses, and peak heights are with criteria windows. Any samples falling outside of these windows must be re-analyzed. Quality control reagent blanks, spikes, and check compounds must be run periodically as in the VOA analysis. A schematic of method 625 semi-VOA analysis is shown in Figure 10.2.

10.3   EPA AND STATE REPORTING REQUIREMENTS

When all samples have been properly analyzed and approved by quality control, the data are used to prepare the data package: BFB or DFTPP tuning reports, sample reports, surrogate reports, internal standard reports, blank reports, and TIC reports, if unknown compound analysis is required. EPA requires that CLP laboratories submit all of these reports along with the raw data in diskette form.

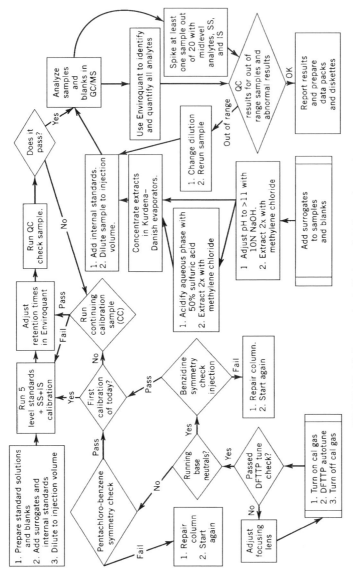

FIGURE 10.2 Method 625 semi-VOA schematic.

The EPA continues to modify their reporting requirements with modifications to the methods occurring ever 3−5 years. Concurrent changes must be made to the quantitation software. Many CLP-like software packages are available, but they do not try to stay current with these changes. Various state and industrial laboratories modify the reporting requirement for their own needs. But many allow much more compact reporting, and generally none are as strict as the EPA requirements.

11

GC/MS IN FORENSICS, TOXICOLOGY, AND SPACE SCIENCE

Environmental analysis has been the most common application of GC/MS systems, but earlier on there had been some application in organic and pesticide manufacturing and in hospital laboratory analysis. The decision to use GC/MS for definitive compound identification in these application areas was often a historical necessity created by the late introduction of the LC/MS systems. Gas chromatography separations works best with nonpolar, volatile compounds, but many of the compounds in nature are polar, reasonably nonvolatile, and thermally degradable.

But, if the only tool you have is a hammer, every problem comes to resemble a nail. GC/MS protocols have often been developed for analysis that might better have been done with LC/MS simply because of availability of equipment. Fragmentation analysis using spectral databases to quickly identify components in a mixture gives GC/MS a definite edge over simple APCI–LC/MS systems that can provide only molecular weight identification of the same compounds. Technological advances in systems miniaturization has once again opened up new application areas for portable GC/MS.

GC/MS: A Practical User's Guide, Second Edition. By Marvin C. McMaster
Copyright © 2008 John Wiley & Sons, Inc.

11.1 FORENSIC ANALYSIS

The human body provides a number of fluids and tissues useful to the forensic laboratory. Blood, urine, semen, sweat, hair, and skin cells are all disposable material that can be extracted to provide information on drugs and poisons that may have been administered to the victim. Postmortem stomach and intestinal contents as well as metabolically active organs such as the liver, spleen, lungs, and kidneys can be examined with GC/MS following tissue excising, homogenization, and extraction to provide information. Hair and fingernails can provide information on chronic poisoning. Comparison of drug or poison concentrations in roots to tips can help in distinguishing long-term chronic poisoning versus acute short-term ingestion.

Analysis of ratios of low molecular weight fatty acids in tissue provides information on bacterial action and can be used in time of death investigations. Bacterial action metabolizes fatting tissue releasing a whole range of short-chain fatty acids immediately after death. Over a period of time, the shortest acids are lost, resulting in a measurable increase in C6–C10 fatty acids easily derivatized and identified by GC/MS.

Crime scene investigation uses GC/MS for identification of drug samples seized at the scene. GC/MS analysis can provide a definite identification of the chemical nature and concentration of the active ingredients and a fingerprint of the cutting agents used to dilute or enhance the action of the intoxicant. Fingerprinting of other components of diluted drugs can be very useful in tracing the drug supply chain back to their source of supply.

Fast scanning protocols are usually run for identifying acidic drugs of abuse such as LSD and barbiturates, basic drugs such as amphetamines, cocaine, and opiates, and specific extractive procedures for THC from marijuana. Designer drugs may appear as unknowns identified by spectra databases in the fast scans, but may require development of specific separation procedure and fragmentation structural analysis. Illegal sport doping agents, such as anabolic steroids in blood and urine, can be identified and quantified in an effort to protect competitive athletics.

11.2 CLINICAL DRUG ANALYSIS

Drug used in hospital and outpatient treatment often have a narrow range between their therapeutic and toxic levels. GC/MS monitoring of blood levels can provide physicians a way of titrating drug levels and preventing overdosing since there is wide difference in metabolic rates and drug detoxification between individuals. Many GC/MS procedure developed for patient studies are being moved over to LC/MS because most therapeutic drugs are water soluble to aid ingestion and distribution in the blood stream. LC/MS analysis usually does not requires sample derivatization and require & simpler and faster extraction methods. Older hospital laboratories tend to hang onto existing GC/MS methods and equipment, while newer laboratories tend to move toward simpler, faster LC/MS equipment and methods. Recent trends of teenagers raiding medicine cabinets for supplies for "buzz parties" see many of these therapeutic drugs moving into the area of drug of abuse and becoming the problem for forensic laboratory analysis.

11.3 ARSON AND SECURITY ANALYSIS

Debris from house and automobile fires is analyzed with GC/MS techniques to look for the presence of accelerants in an attempt to prove that arson was responsible for the fires. Residues of gasoline, diesel fuel, kerosene, fuel oil, and other flammable solvents can be extracted from charcoal, ash, auto fabric, and wood residues. Explosives, such as dynamite, primers, and plastique, used as accelerants or fire initiators are never completely consumed and can be extracted and identified by high sensitivity GC/MS or LC/MS analysis.

With security concerns on the increase, dedicated headspace GC/MS systems are finding growing application as "bomb sniffers" for detection of flammable solvents and explosives in packages and luggage. While still not as effective as the nose of a trained dog, the sensitivity offered by the mass spectrometer in detecting and definitively identifying the nature of the accelerant offers real promise to retire the canine officers in the not too distant future.

11.4 ASTROCHEMISTRY

Advancing techniques in GC/MS miniaturization have allowed this technique to move into space exploration. Every planet-landing explorer has carried GC/MS systems to analyze atmospheric gases and remote laboratory systems to acquire and process terrain samples to examine for organic chemicals indicative of extraterrestrial life.

The two Mars Viking landers were both equipped with GC/MS laboratories. The first stationary lander provided atmospheric gas compositions, but the only surface organic components identified turn out to be the result of landing exhaust gases. The very successful, more sophisticated crawler on the second Viking lander was able to sample atmospheric gases and soil samples from a number of locations, but added little to the inventory of Martian organic chemicals.

The three Venus landers, Russian-built Venera 11 and Venera 12, and the American Pioneer Venus, all carried GC/MS systems to analyze atmospheric gases on the way in. Crash landings and the extremely high surface temperature and toxic atmosphere prevented any sampling for surface chemicals.

The most recent lander system is the Huygens probe parachuted by the Saturn-orbiting Cassini spacecraft onto the moon Titian on January 14, 2004. The probe carried a triple-column GC/MS system to analyze atmospheric gases on descent and continued to send information to the Cassini spacecraft from the ground for 71 min until the orbiter moved over the horizon and lost contact with the probe.

The next planned GC/MS space analyzer mission is planned for 2014. The Rosetta mission will send a spacecraft to rendezvous with the comet 67P/ Churyumov−Gerasimenko where its chiral GC/MS will attempt to analyze the composition of the comet's tail gases.

12

AN INTRODUCTION TO STRUCTURAL INTERPRETATION

I make no claims to being an expert in structural interpretation of mass spectra. This technique is a science in itself and beyond the scope of this book. The information presented in this chapter is presented as a guide to the way I have used this technique in unraveling some questionable library assignments.

Interpretation of molecular structures from fragmentation data is an involved, time-consuming, and exacting science. If you are the type of person who enjoys doing *The New York Times* crossword puzzles, you might find it worth pursuing in more detail. Drs. McLafferty and Throck-Watson have excellent books listed in Appendix E designed to help you learn to extract structural information using ion fragmentation mechanisms. Table 12.1 summarizes the main points I found and the order in which you need to acquire information from the spectra.

When running a GC/MS system in the laboratory, you will probably never need or want to do a rigorous interpretation of a structure. It is faster and a far better use of your time to do a library database search and let the computer match your fragmentation pattern to known spectra. Structural interpretation is of great value, however, in confirming the structural assignments found by the computer database search.

GC/MS: A Practical User's Guide, Second Edition. By Marvin C. McMaster
Copyright © 2008 John Wiley & Sons, Inc.

TABLE 12.1   A Guide to Molecular Structure by Fragment Analysis

1. Base peaks and relative ion intensities.
   (a) Determine molecular ion mass. Run CI if needed.
   (b) A scarcity of major low even-mass ions = an even-mass MW.
2. Elemental composition from isotopic abundances.
   (a) Look for $A + 2$ pattern elements (Cl, Br, S, Si, O). (Check $A + 1$ ratios
       for absence/presence of S, Si.).
   (b) Use the nitrogen rule to determine no. of nitrogens. (If MW is even = 0 or even
       no. of N. If MW is odd = odd no. of N.).
   (c) Number of carbons/nitrogens from $A + 1$ isotopic ratios.
   (d) Estimate no. of H, F, I, P from A isotopic ratios and MW balance.
       (Only P is multivalent. F = 19 and I = 127 mass units.)
   (e) Check allowance for rings and double bonds. (No double bond or
       rings $= x - 1/2(y) + 1/2(z) + 1$.) where $(C, Si)_x(H, F, Br, Cl)_y(N, P)_z(O, S)$.
3. Use molecular ion fragmentation mechanisms.
   (a) Check fragment masses differences for expected losses. (35 = Cl,
       79 = Br, 15 = Me, 29 = Et, etc.)
   (b) Look for expected substructures.
   (c) Look for stable neutral loss ($CH_2=CHR$).
   (d) Look for products of known rearrangements.
4. Postulate structures.
   (a) Search library database.
   (b) Run hit compound on same instrument to confirm.
5. Use MS/MS if further confirmation is needed.

One of the problems with spectral library databases is that some of their structures are inaccurate or just plain wrong. The original interpretation of their structures may have been incorrect or mistakes may have been made in entering them. The previous Wiley spectral database with 225,000 compounds was thought to have up to 8% incorrect structures. It is claimed that the current Wiley database has been cleaned up and that the structure assignments are >98% accurate.

Even when you are working with accurate known spectra and precise spectral matching algorithms, there are still sources of problems:

1. The tuning conditions use in preparing the target spectra may not have been the same as those used in the laboratory.
2. The spectra could have been run on a different type of mass spectrometer with a different mass linearity. Some data in these libraries were run on magnetic sector instruments rather than on a quadrupole. The high mass areas of these two types of instruments do not calibrate the same ways.

3. Either your spectra or the target spectra may have been run on impure compounds, which may introduce additional fragment peaks, especially at low relative intensities which may affect matching.

4. You may have chosen to scan above 40 amu to avoid water and air peaks, while the target spectra may include these extra fragments, altering the match.

The library matches do not give you single compounds, they provide you a list of possible matching compounds with a percent confidence level for that particular match. You may have a pair or more of possible structures between which you will be required to choose. Partial structure interpretation can be a useful guide to making choice between close matches or in determining whether a high probability match makes any sense at all.

12.1 HISTORY OF THE SAMPLE

The starting point for examining a fragmentation spectrum is to find out as much as possible about the compound being examined. Where did it come from? What kind of solubilities does it show? Does its UV spectrum show conjugation or aromatic structures? What can the chromatography tell you about its polarity? Does it show hydrogen bonding when run under conditions that break such bonds? What kind of compounds does it separate with? What is its boiling point or melting point? What is its molecular weight? The more you know about the compound before you start your separation, the quicker you will be able to confirm the apparent match from the database.

Once we have its molecular weight and an idea of its chemical nature, we can move on to determine its elemental composition from isotopic abundance information calculated from fragment patterns. Finally, we examine the mass differences between ion fragments to determine what types of groups are being lost. If we do at least this much, we almost always have enough information to confirm a library structure assignment.

Molecular weight information is available from the compound's mass spectrum. If the molecular ion is missing from the fragmentation pattern, being so unstable that it contributes at best only a very tiny peak, we can switch over and run in the chemical ionization mode. That will give us a molecular ion and, therefore, the compound's molecular weight as the first piece of information we must have to begin our analysis. The fragmentation pattern will point us toward whether to expect an even or an odd mass molecular weight. If we look through our pattern and see a scarcity of major even-mass ions in the low mass range, we probably have an even-mass molecular weight.

12.2   ELEMENTAL COMPOSITION

Next we need to determine how many carbon, hydrogen, nitrogen, oxygen, and other elements are present. We can do this by looking for elements that show characteristic isotopic patterns in the fragment spectrum. Try to work with the most massive fragments and with the fragments having the tallest mass peaks. In any group, start with the most intense ion fragment in the group, the one with the most stable isotope, as the M peak. Smaller adjacent ion peaks, such as $M + 1$ and $M + 2$, represent isotopic relatives of the A ion fragment.

Table 12.2 is a list of isotopic ratios for common elements making up organic molecules. A fragment containing an A-type element shows only a single band in the spectra. An $A + 1$ element, such as carbon or nitrogen, has two isotopic forms separated by 1 amu and forms pairs of fragment ions. The relative intensity peaks of the fragments will be the same as the isotopic abundance of the element. If an ion fragment has a single carbon, the relative height of the first mass peak would be 100; 1 amu up would be a fragment with a height of 1.1. The effect is additive. The more carbon atoms in the ion fragment, the higher will be the $M + 1$ peak height, that is, if five carbon atoms are present, the second peak would have a height equal to $5-6\%$ relative to the first peak. If you do nothing else, find these $A + 1$ fragment pairs and use them to estimate the carbons present in each fragment. When you are working with organic molecules you will be right most of the time. Biological molecules have enough nitrogen molecules to throw this number off.

TABLE 12.2   Natural Isotopic Abundances of Common Elements

Element	A Mass	A %	A + 1 Mass	A + 1 %	A + 2 Mass	A + 2 %	Element type
H	1	100	2	0.015			A
F	19	100					A
P	31	100					A
I	127	100					A
C	12	100	13	1.1			$A + 1$
N	14	100	15	0.37			$A + 1$
O	16	100	17	0.04	18	0.2	$A + 2$
Si	28	100	29	5.1	30	3.4	$A + 2$
S	32	100	33	0.79	34	4.4	$A + 2$
Cl	35	100			37	32.0	$A + 2$
Br	127	100			81	97.3	$A + 2$

Before we can continue to work on carbon, nitrogen, and hydrogen, we must first determine the presence or absence of other elements. Fragments containing the so-called $A + 2$ elements show a large A peak and 2 amu higher a smaller peak of a precise height depending on which element you are seeing. Chlorine stands out like a sore thumb. A primary fragment peak containing chlorine shows an $M + 2$ secondary fragment one-third the height of the primary fragment. Every fragment containing only a single chlorine will show this same $3 : 1$ $A + 2$ ratio. This pattern occurs because chloride is a mixture of isotopes, its major isotope has mass 35, but it has a second major isotope with mass 37 with 32% of the 35 mass isotopic abundance. You can tell when an ion fragment decays with loss of this chlorine. The mass difference between fragments will be 35 and the $A + 2$ pattern will not appear in the smaller fragment. Bromine shows an $A + 2$ doublet of almost equal height (100% and 97%). Compounds with multiple chlorine molecules or a mix of chlorines and bromines in the same molecule show other characteristic patterns that are the additive results of combining $A + 2$ patterns. Tables of these are available in Throck-Watson's book (see Appendix E).

Once we have determined the number of chlorines and bromines present and subtracted their contributions to the molecular weight, we need to look for the presence of sulfur and silicon. These also are $A + 2$ elements, but they show an additional isotope at the $M + 1$ position. First you find $A + 2$ patterns, then look for a mass in the middle. If there is no intermediate peak, you can scratch off these two elements. If there is an intermediate $M + 1$ peak, compare its height to the M peak height after you have removed any chlorine and bromine contributions and compare it to the values in Table 11.2. This should lead you to the number of sulfurs, or less likely the number of silicon, present. Oxygen is also an $A + 2$ element, but the isotopic contribution from C^{18} is too low to be useful for measuring the amount of oxygen present in a fragment. Usually, it is estimated from the residual molecular mass after the other elements except H are eliminated.

Once we have eliminated the $A + 1$ contributions from sulfur and silicon, we are ready to calculate the carbon and nitrogen values. In a simple hydrocarbon such as hexane, we should expect the $+1$ fragment peak to be about 6.6% as high as the main peak. Contributions by nitrogen are estimated using the Nitrogen Rule. This states that if the molecular weight is even, you will see either no nitrogen or an even number of nitrogens in the fragment. Odd molecular weights occur when there is an odd number of nitrogens. This allows us to either eliminate nitrogen or to come up with a satisfactory number of nitrogens. Subtracting the nitrogen contribution should provide us with a good ratio of carbon isotopes, allowing us to calculate the number of carbons present. We can now subtract the carbon and nitrogen contributions to the molecular weight.

We are now left with hydrogen, fluorine, iodine, phosphorous, and, of course, oxygen. Because of its large isotopic mass, the presence or absence of iodine is usually obvious at this point, and can usually be eliminated. Phosphorous is multivalent and most commonly bound to multiple oxygens, it is usually easy to eliminate or identify. Fluorine's odd mass of 19 and its univalent replacement of hydrogen makes its presence or absence apparent when you are trying to distribute the residual molecular weight units between oxygen, hydrogen, and fluorine. Once you have an elemental assignment in hand or even a partial that makes sense, check whether it agrees with the compound selected by the library search engine.

One more check that can be done is to check the number of double bond and rings that are present. Table 12.1 presents a formula for calculating this number. You add up the number of quadruvalent, trivalent, divalent, and monovalent atoms present and plug them into the formula. You end up with a number representing the total number of double bonds and rings present using the lowest valence state for the elements. For a benzene ring, this number would be 4, for an electron balanced, charged ion this number might be 1/2.

12.3   SEARCH FOR LOGICAL FRAGMENTATION INTERVALS

The final thing I look for in a spectra is mass losses between major fragment peaks. I look for characteristic losses like 35 (CL−), 15 (CH$_3$−), 29 (CH$_3$CH$_2$−), or a 15 loss followed by a series of 14 (−CH$_2$−), which indicates a breakdown of a straight chain hydrocarbon. Also look for neutral molecule losses such as substituted vinyls (RCH=CH$_2$), which occur as part of rearrangements, and 28 (carbon monoxide), which may indicate the presence of a carboxylic acid or an aldehyde.

Once you find these markers, go back to the library structures and see if you can tell where these pieces are coming from. If none of these breakdowns make any sense, you may not have the right structure. If you can see how the pieces you are seeing can be formed, you have found additional confirmation for the structural assignment.

I hope this makes sense and helps you in confirming assigned structures. A rigorous study of fragmentation mechanism will let you recognize many more loss assignments, but you will have to determine whether it is worth your time.

In any case, the ultimate test is to acquire a sample of what you believe to be the correct compound. Run it on your GC/MS system with your tuning parameters under your chromatographic conditions to see if it gives the same spectrum.

13

ION TRAP GC/MS SYSTEMS

Ion trap mass spectrometers (ITDs, ITMSs, and LITs) are finding growing acceptance in GC/MS laboratories. Laboratories that use them claim they are 10–100 times more sensitive than a quadrupole. They can easily be switched between CI and EI modes, require less maintenance, and have potential to be used for MS/MS studies, especially in investigation of trace contaminates.

The desktop ion trap detector (ITD) and the floor-standing ion trap mass spectrometer (ITMS) vary in size and added function more than in theory of operation. The ITMS is designed as a research instrument with both analytical MS and MS/MS operation in mind. The ITD is a dedicated, compact unit with a smaller trap and pumping system designed for production GC/MS operation. The newest member of the ion trap family is the linear ion trap (LIT), which is essentially a quadrupole with electrical lens at both ends to hold ions within the detector body.

Molecules introduced into the ion trap are processed totally within the body of the ITD ion trap. Uncharged material from the GC stream enter the trap around the ring electrode, are ionized, collide with other molecules, fragment, and is stored in stable orbits between the electrodes. The stored ions are then eluted in increasing mass (m/z) by increasing the voltage on the ring electrode. This pushes each fragment ion into an unstable orbit, causing it to escape through one of the seven holes in the exit electrode and into the dynode

GC/MS: A Practical User's Guide, *Second Edition*. By Marvin C. McMaster
Copyright © 2008 John Wiley & Sons, Inc.

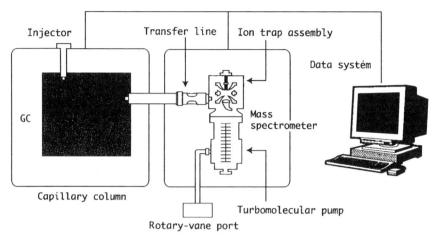

FIGURE 13.1 Ion trap GC/MS system.

electron multiplier detector, which sends a signal to the data system (Fig. 13.1).

13.1   ION TRAP COMPONENTS

The ITD GC/MS system is contained within two connected modules. The GC oven with column, injector, and transfer interface are similar to those used in the quadrupole system. The connection from the interface enters the ion trap through a transfer line just underneath the ring electrode. The detector horn lies immediately below the exit electrode.

The turbo pump is mounted directly below and attached by a vitron O-ring gasket to the ion trap body, both of which are enclosed in a heated vacuum manifold. Also on the manifold are the attachments for the cal gas valve, an entrance line for chemical ionization gas, and an exhaust port for the rotary-vane mechanical pump.

13.2   ION TRAP OPERATION

Only a limited volume of sample can enter the ion trap with out overloading and causing performance degradation. A narrow-bore capillary column with flows of 1 ml/min can be directly interfaced or a splitter column can be used to divert part of the GC stream to a secondary detector. Once the sample is in the trap, it is ionized with 70 V electrons from the ion gate in the entrance electrode at the top of the trap (Fig. 13.2).

Thermionic electrons are furnished by a heated filament. Between the filament and an unused spare filament is a repeller plate that drives the electrons toward the ion trap containment space (Fig. 13.3).

At the base of the ionization electrode is a variably charged electron gate. When the gate has a high negative charge, electrons stay in the electrode; when the gate drop positive, electrons are forced into the storage space and ionize molecules of the sample.

The ring electrode around the containment space has a constant frequency, and a variable amplitude radio frequency signal is applied to it. A storage voltage of 125 DAC is applied to trap all ions with mass equal or greater than 20 amu. At this voltage, the ions formed are thrown into circular, hyperbolic orbits that are described as resembling the stitching on a baseball (Fig. 13.4).

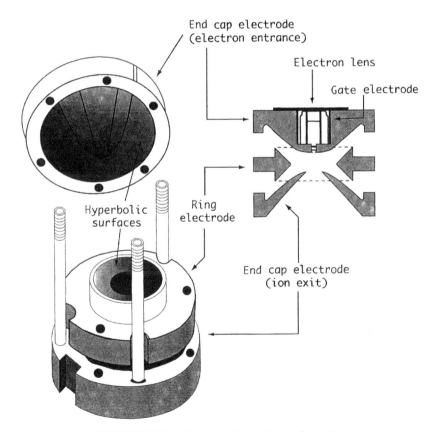

FIGURE 13.2   Ion trap electrode configuration.

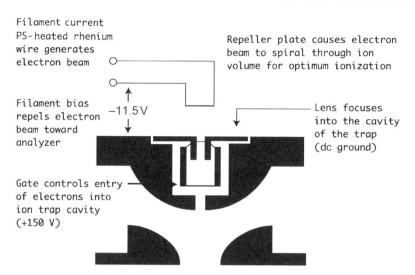

Filament current
PS-heated rhenium
wire generates
electron beam

Repeller plate causes electron
beam to spiral through ion
volume for optimum ionization

Filament bias
repels electron
beam toward
analyzer

−11.5 V

Lens focuses
into the cavity
of the trap
(dc ground)

Gate controls entry
of electrons into
ion trap cavity
(+150 V)

FIGURE 13.3   Ion trap filament and ion gate.

Approximately 50% of all ions formed are thought to be trapped and
eventually reach the detector. This is compared to the single ion at a given
time point that reaches the detector in the quadrupole. Most ions end up
colliding with the quadrupole rods and are never analyzed. This increased ion
yield explains the increased sensitivity of the ion trap. Some increase in ion
trap analyzer stability comes from the lack of sample accumulation on the

FIGURE 13.4   Stable ion trajectory schematic.

electrodes, although this will vary from sample to sample. The helium carrier gas in the GC stream serves an important role in stabilizing the ions in their orbits. Frequent collisions between the small, fast-moving gas molecules and the charged ions dampens their movement, causing them to collapse toward the center of the trap.

The analysis is performed by gradually increasing the ring electrode RF voltage or by scanning the voltage. This upsets the orbits of ions with increasing masses, causing them to escape through the exit electrode's holes and impact on the dynode's surface.

Ion orbital stability is also improved by applying axial modulation. This is a fixed frequency and amplitude voltage applied between the ionization electrode and the exit electrode at a frequency equal to about half that of the ring electrode. It has the effect of moving ions away from the center of the trap where the voltage is zero. This aids in ion ejection from the trap and dramatically sharpens the mass resolution at the detector.

Scanning is done in four segments over the full scanning range. This allows for mass peak–height manipulation and tune modification. With this tool, the tune can be adjusted to meet specific peak ratios requirements to match tunes done on quadrupole systems providing better fit with mass spectral data libraries.

The ion trap detector is the cascade dynode electron multiplier we have previously seen used in quadrupole systems (Fig. 13.5).

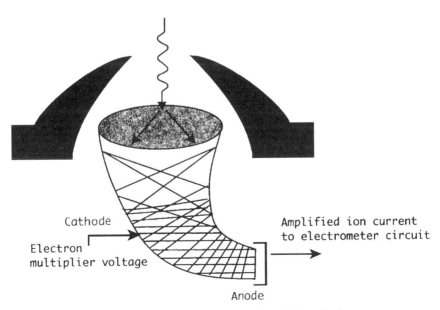

FIGURE 13.5   Ion trap exit electrode and dynode detector.

Positive ions striking the lead oxide glass cathode surface release electrons from its inner surface. These bounce down the inner walls, releasing a cascade of electrons on each contact; as many as 100,000 from a single positive contact will reach the anode cup and send a signal to the data system.

Ion traps are being used extensively in environmental production laboratories. Their high sensitivity and easy maintenance makes them attractive by avoiding downtime and providing trace analysis capability.

13.3   THE LINEAR ION TRAP ANALYZER

The latest hot analyzer is the LIT or linear ion trap. It combines the separating capability of the quadrupole analyzer with the MS/MS capability of the ion trap. Trapping electrode rings are added to each end of the quadrupole rods to create the linear ion trap (Fig. 13.6).

The analyzer can be run in a normal scanning quadrupole mode for separation and detection of mass ions, or the end electrodes can be turned on to retain a specific ion in the trap for collision with a damping gas and fragmentation that may be aided with a supplemental resonance excitation voltage. The daughter ion fragments can then be sequentially released to the ion detector by scanning dc/RF voltage on the quadrupole rods while utilizing a supplemental resonance ejection voltage on the trapping electrodes. The major advantage of the linear ion trap over the circular ion trap is the capacity of the linear ion trap. A normal ion trap is a point source trapping ions in a spherical segment between the ring electrodes. A linear trap spreads the sausage-shaped trapping volume down the center of the quadrupole pole rods greatly increasing the trap's capacity. Reports in product brochures and the literature claim this increase to be from 10- to 70-fold than that of the circular ion trap. This translates in an increase in sensitivity for analyzing minor components of the HPLC effluent for trace analysis of metabolites or minor fragments from protein sequencing. Current linear ion traps are expensive,

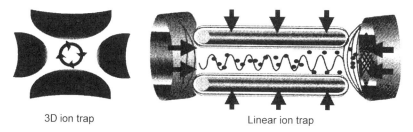

3D ion trap   Linear ion trap

FIGURE 13.6   Linear ion trap analyzer. (Courtesy of BioAnalytical Systems.)

free-standing research instruments, but refinement and simplification of the technique seems to offer great potential for producing an inexpensive, high sensitivity desktop GC/MS/MS.

13.4   ION TRAPS IN THE ENVIRONMENTAL LABORATORY

Ion traps have received a bad rap in the past in that they do not produce spectra that match existing spectral libraries. This seems to have come about from poor autotune software in earlier traps. Using the four segment wavelength scanning to balance tunes, they are able to meet environmental tuning parameters for BFB or DFTPP analysis. They yield spectra that are easily searched and identified from either Wiley or NIST libraries. Libraries of ion traps only data are starting to emerge with specific scanning parameters for ion traps, but these may prove useless wherever tuning compounds and parameters have been defined and reported.

13.5   CHEMICAL IONIZATION IN THE ION TRAP

One of the biggest, but least appreciated advantages of the ion traps is its ability to run chemical ionization without switching the ionization source. Since ionization occurs in the trap itself it is only necessary to introduce a primary ionizing gas such as methane, butane, or ammonia. Since the helium pressure is already around 10^{-3} and circular ion traps are very susceptible to overloading, some loss of sensitivity may occur due to space charging of the trap volume. But switching is so easy that it is possible to time program the jump from EI to CI in the same run. Since CI provides us with the molecular ion mass, it is a valuable aid in determining structure for an unknown compound when combined with the fragment information from the EI run. The higher storage volume of the linear ion traps avoids the overloading problem seen with the circular ion traps when running in a CI mode.

13.6   ION TRAP GC/MS/MS

Gas chromatography/MS/MS mass spectrometry is possible in an ion trap because alternate waveforms can be used to store specific selected ions in the trap. By allowing these ions to collide either with themselves or with a heavy makeup gas ion, such as xenon, the ion will fragment. The daughter ions produced can be used to help identify the parent ion and, by examination of fragmentations of a series of fragments, to identify related fragments from the original breakdown. MS/MS will be covered in more detail in the next chapter.

14

OTHER GC/MS SYSTEMS

While quadrupole and ion trap systems represent the majority of commercial systems sold and used, other GC/MS systems are coming into use for specific purposes. The first GC/MS systems were magnetic sector system that generated an electromagnetic field to deflect the flight path of charged molecules in to curves. Due to their unusually stable magnetic fields, these system still have application in very accurate determination of molecular weights. Triple-quadrupole and hybrid GC/MS/MS systems are the favorite tools of research and method development departments for determining structures of primary fragments. Time-of-flight (TOF) mass spectrometer systems have some general applications for volatile molecules, but show growing use in LC/TOF-MS for molecular weight determination of large molecules such as proteins and DNA restriction fragments. The Fourier transform mass spectrometer used in GC/FTMS is a research tool offering much faster accurate mass determination and higher sensitivity for trace component analysis than other existing techniques.

GC/MS: A Practical User's Guide, Second Edition. By Marvin C. McMaster
Copyright © 2008 John Wiley & Sons, Inc.

14.1   SEQUENTIAL MASS SPECTROMETRY (TRIPLE-QUADRUPOLE OR TANDEM GC/MS)

The so-called triple-quadrupole mass spectrometer is in reality an instrument made up of two scanning analyzers separated by a collision cell. Fragments selected in the first quadrupole collide with inert gas, usually a large molecule like Xenon, in the center quadrupole and undergo further fragmentation. The secondary fragments are then resolved in the final quadrupole analyzer (Fig. 14.1).

The ionization source, focusing lens, and detector sections are identical to those in a single-quadrupole system. Collimator lens after the collision cell focus the secondary ion fragments into the second analyzer. The purpose of the triple quadrupole is to allow separation and selection of primary fragments in the first analyzer, fragmentation of the separated primaries in the collision cell, and analysis of secondary fragments in the second analyzer.

There are four possible modes of operation of the two analyzers: Q1, SCAN/Q3, SIM called "daughter mode"; Q1, SIM/Q3, SCAN called "parent mode"; Q1, SCAN/Q3, SCAN referred to as "neutral loss scanning mode"; and Q1, SIM/Q3, SIM referred to as "MRM, multiple reaction monitoring mode" (Fig. 14.2).

The SCAN/SIM mode operation lets us determine which primary fragments are related to each other. The first quadrupole is scanned over a mass range and all fragments formed enter the collision cell and fragment to form secondary fragments. The third quadruple is parked at a specific mass; only primary fragments that break down to form this specific secondary ion will be detected. This common daughter ion points out interrelated primary fragments and helps us to understand which fragments are formed when a large primary fragment breaks down.

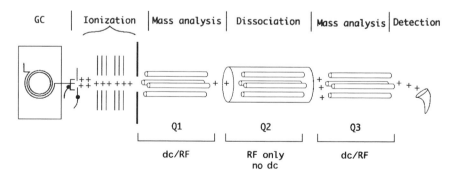

FIGURE 14.1   Triple-quadrupole GC/MS system.

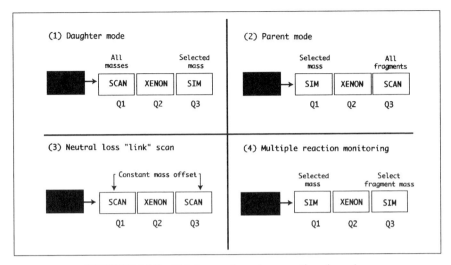

FIGURE 14.2   Triple-quadrupole operational modes.

The SIM/SCAN operations parks the first quadrupole analyzer at a specific mass allowing only a single primary fragment to enter the collision cell where it fragments into secondaries. The final quadrupole is run in full scan mode detecting all secondary fragments formed from this single primary parent, again providing structural information by showing its breakdown products.

The SCAN/SCAN operation is a little more complicated since both analyzer quadrupoles will be scanned at the same time, but with a predetermined mass offset. When a primary fragment under goes further fragmentation, it breaks into two pieces, a charged secondary fragment and a neutral molecule. What we are detecting in this mode are primaries that lose the same neutral molecule and therefore may be breaking down by the same fragmentation mode. The molecular mass of our suspected "neutral loss" is the value we assign to our scan offset between the two quadrupoles. All primary fragments separated in the first analyzer enter the collision cell and fragment. Only secondary fragments whose mass is exactly the neutral loss smaller than their primary fragment are detected by the final quadrupole. The second analyzer will not detect any primary fragment that breaks down by forming a neutral molecule having a mass different from the offset mass.

The SIM/SIM operation is designed to definitively analyze specific components of very impure mixtures without having to completely purify them. Nature makes very complex mixtures that cannot always be completly separated either though extractions or by chromatography. We examine a chromatographic peak in which we expect a specific compound to appear by using the first quadrupole to separate a primary fragment characteristic of the compound of interest, pass it into the collision cell, and use the final quadrupole to identify it by looking for only one of its specific daughter ions. We can identify and quantitate each targeted compound in a mixture, even if the chromatographic peaks that contain them are contaminated. For each compound to be analyzed, we select an individual primary and secondary fragment on a time basis in step with their expected chromatographic retention time.

Not all MS/MS systems use dual quadrupole analyzers. As we mentioned in the last chapter, ion trap systems have built in MS/MS capability. An ion trap/quadrupole system can be designed to pass specific parent ions to a quadrupole for daughter ion analysis. Or a specific daughter from ion trap MS/MS can be passed through the collision cell for quadrupole fragmentation analysis. Ion trap or quadrupole/TOF MS/MS analyzers use the first analyzer to act as a storage cell to feed the time-of-flight analyzer that operates in a burst mode. Modern quadrupole/magnetic sector MS/MS combines the capabilities of both types of mass spectrometers, rapid scanning of the quadrupole for fragment separation and the stability of the magnetic sector secondary for accurate analysis of daughter ions. The literature contains examples of even more complex research systems: MS/MS/MS and even more exotics MS^n hybrid separating mode, multiple-analyzer systems, but these research tools are beyond the consideration of a practical book on GC/MS systems.

14.2   MAGNETIC SECTOR SYSTEMS

The first research and commercially available GC/MS systems were magnetic sector instruments. They use an electromagnet based on a large permanent magnet to force ion fragments into circular sector flight patterns whose curvature is dependent on the fragments mass/charge ratios (Fig. 14.3).

The lighter the m/z mass, the more the deflection that it exhibits in the magnetic sector. Scanning of the mass range of ion fragments can be achieved in one of two ways, either by varying the accelerating voltage in the source or by scanning the electromagnetic field strength of the magnetic sector. Detection is done by using a moving slit and a photomultiplier tube or by an electro-optical linear array.

The limitations of the magnetic sector systems are cost, the size and weight of the permanent magnet, response time, sensitivity, and linearity, especially at the high mass side. The spectra obtained are generally not directly comparable to results from quadrupole or ion trap instruments for spectral extraction and library searching. Variation of the accelerating potential is

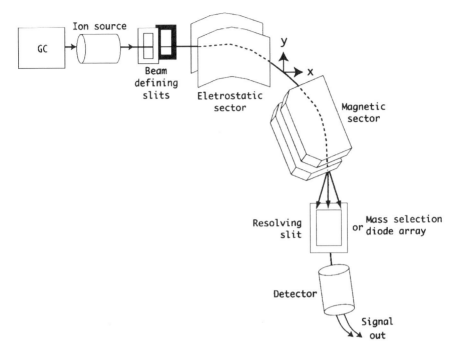

FIGURE 14.3   Magnet sector mass spectrometer.

limited in mass range and sensitivity drops off at the high mass end. Scanning the magnetic field is the more commonly used technique, but it suffers from reluctance, an inertia resistance to magnetic field change. This leads to slower scan rates that translates to poor sensitivity. Much of this sensitivity problem is overcome in modern instruments by employing spatial-array detectors so that the whole mass range can be measured at all times, increasing the sampling rate and efficiency.

Magnetic sector instruments have made a comeback in the last few years because of their importance in accurate mass measurements for precise molecular weight determination. Injection is made from a probe rather than a gas chromatograph. A technique called peak matching is used to compare the difference in accelerating voltage need to make an unknown and a reference ion reach the detector in coincidence. The reference ion must be within 10% of the unknown compound mass, and masses can be measured to six-decimal place accuracy with this technique.

Double-sector instruments are used to increase the precision by using in series an electric sector to select ions of only one specific kinetic energy and the magnetic sector to peak match the reference and unknown compounds. Double sectors can be used for isotopic mass determination; they separate all ionic species into separate peaks which can be peak matched against a reference compound.

14.3   LASER TIME-OF-FLIGHT (GC/TOF-MS) GC/MS SYSTEMS

A growing segment of the GC/MS system market is using GC/TOF-MS systems. The TOF mass spectrometer uses a 3-kV electron beam to burst ionize the sample from the GC in the mass spectrometer source. The fragments are repelled down a flight tube through focusing lens.

The flight time of each fragment is dependent on its m/z ratio, lighter fragments arriving first at the detector. To detect a given m/z fragment, the electron-multiplier tube is activated only for a given time slice window allowing selection of only a single mass per burst. Flight time is very rapid, on the order of 90 ns for a 2-m flight tube. By stepping the time window for the electron multiplier for subsequent burst, all masses over a range can be sampled and averaged fast enough to detect and analyze the narrow peaks produced by a gas chromatography.

Since the majority of the fragments from every burst are discarded, sensitivity and resolution are a potential problem. SIM mode operation is the natural operating mode for a TOF instrument since the electron-multiplier time window does not have to be stepped. To increase sensitivity, a timed array detector is used in newer instruments. The array elements are set to sample the flight stream reaching the detector at different time windows. Using this technique, the whole burst fragment pattern can be analyzed for each event. Summing the resulting time windows allows a 10,000-fold increase in sensitivity. Arrays are limited by the number of array elements available to do the sampling and by the inherent noisiness of the array. A 50×50 array provides 2500 sample points. For a $0-800$ amu detection range, this provides a 0.3 amu resolution. Typical quadrupole resolution is 0.1 amu or better.

The length of the flight tube has historically produced very large, cumbersome TOF instruments. Folding the tube using electrical "mirrors" to reflect and accelerate the fragment flight stream back down the flight tube to impact the detector has greatly reduced this problem, see Figure 14.4.

Time-of-flight GC/MS systems are rare outside academia. This MS technique is having more success in LC/MS where LC/MALDI-TOF/MS systems are used for analysis of proteins, peptides, and polynucleotides. (MALDI is short for maser-assisted laser desorption and ionization). The liquid stream from the HPLC is mixed with a chromaphore, such as cyanocrotonic acid, that will absorb light from the high intensity laser burst in

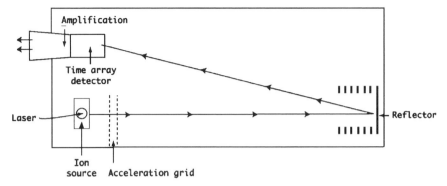

FIGURE 14.4 Reflectron time-of-flight GC/MS system.

the source. These target dye molecules explode throwing the accompanying protein into the gaseous phase and at the same time chemically ionizing it. Since the free amino groups on side chains provide multiple ionization sites, a series of multiple charged molecular ions each with a different charge are formed from a single protein. These are repelled down the flight tube and separated at their m/z masses. Analysis of this family of molecular ions, which differ by the size of their charge z, allows calculation of the molecular weight of the original protein. Charges as large as $20-50$ on an ion radical allow enzymatic size proteins (MW $= 25-60$ kDa) to be separated on a TOF mass spectrometer with a separating range of $0-1000$ amu.

14.4 FOURIER TRANSFORM (GC/FT-MS) GC/MS SYSTEMS

Gas chromatography/Fourier transform mass spectroscopy (GC/FT-MS), produces mass spectra using ion cyclotron resonance (ICR). The sample is ionized by a burst of electrons in the source and passed into the analysis cell where they are held in a constant magnetic field provided by trapping plates. Each fragment will follow a circular orbit with a cyclotronic frequency characteristic of its m/z value.

To detect the fragments present, a full frequency RF "chirp" signal is applied from a transmitter plate perpendicular to the trapping plate. Ions absorb energy from the chirp at their cyclotronic frequency and are promoted to a higher orbit. Detector plates perpendicular to the third plane of the cell measure a complex signal containing all the frequencies of the excited fragments (Fig. 14.5).

Fourier transformation software converts this frequency snapshot to a spectrum of the m/z values present in the sample. Like the spatial array detectors,

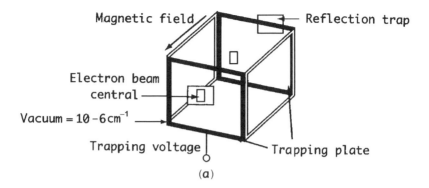

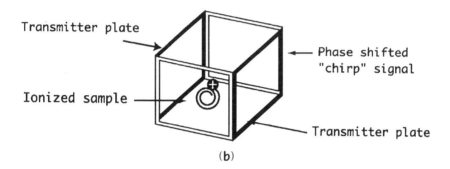

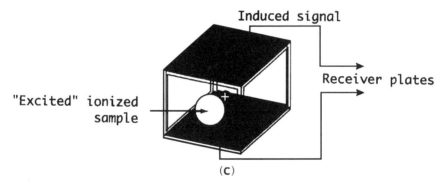

FIGURE 14.5   Fourier transform GC/MS system. (**a**) Pulse. (**b**) "Chirp" excitation. (**c**) resonate frequency signal.

every fragment is analyzed for every ionization burst event. Ionization can be in the equivalent of electron ionization (EI), chemical ionization (CI), and laser-assisted ionisation, the examples of which exist in the literature.

Measurements can be made in milliseconds and have been used to monitor very short gas-phase reactions. Since the ion fragments are not destroyed in the cell, multiple measurements over time can be averaged to produce a very accurate, high resolution measurement yielding excellent sensitivity. The signal tends to be very stable and is not dependent on ion optics or variation in detector electronics. Modern computers can provide transformation calculation fast enough to provide real-time data. This analyzer offers great promise for accurate mass determination and trace component analysis and as a component of an inexpensive general use GC/MS system once price, technology, and computing power issues are addressed.

15

AN INTRODUCTION TO LC/MS

Interfacing a mass spectrometer to a high performance liquid chromatography, an important potential addition to the HPLC arsenal, is not a new technique. The mass spectrometer began as an out growth of the Manhattan Project during World War II. As investigators involved in this program returned to their respective universities, the techniques they had developed, and in many cases, the equipment returned with them.

In the 1960s, a GC/MS interface was developed, but the first HPLC/MS interface did not appear until the 1970s because of the problem of detecting compounds in the presence of all that solvent. The mass spectrometer is very nearly the perfect HPLC detector since it allows non controversial identification of even "unknown" compounds from their fragmentation spectra. The problems preventing widespread introduction of LC/MS into general laboratory use have been threefold: the price, getting rid of large amounts of HPLC solvent, and expertise in interpreting results. Ten years ago, an LC/MS system was a massive instrument costing in excess of $100,000 for the mass spectrometer, an LC Interface costing in excess of $20,000, and the $30,000 price of a gradient HPLC. The high vacuum pumps required to run the system needed constant maintenance. The high level of organic solvents in the HPLC mobile phase tended to overload the mass spectrometer and overwork the

GC/MS: A Practical User's Guide, Second Edition. By Marvin C. McMaster
Copyright © 2008 John Wiley & Sons, Inc.

pumping systems. Interpretation of spectra was tricky and required someone specialized in the field.

Many of these problems are rapidly disappearing. Desktop mass spectrometry detectors (MSD) have shrunk in size and prices have dropped to around $50,000. Pumping systems are becoming less demanding of service. The solvent problem is finely coming under control in atmospheric pressure interfaces using nebulizers, heaters, and splitters. Interpretation of results is much easier and faster due to automated mass identification and computerized online, rapid spectral library database searching of fragmentation data.

While prices of new systems are still prohibitive for the average chromatography laboratory, older systems have been retrofitted with modern data systems, equipped with home-built LC interfaces and mass spectrometers recovered from GC/MS systems and put back into operation for around $30,000. As prices dropped and technology advanced, the LC/MS has become a major tool for the clinical chemist measuring blood levels of therapeutic drug, forensic chemist, analyzing drug of abuse, and arson investigator, analyzing the presence of accelerants at a fire scene, whose data must stand up in court. The food and environmental chemists must analyze samples that effect the food we eat, the water we drink, and the air we breathe. All analyze a broader range of material more rapidly without the limitation provided by the gas chromatography. Almost any compound that will dissolve can be separated in an HPLC, and potentially, analyzed by a mass spectrometer.

Let us take a look at the design of the LC/MS, its operation, and the way mass spectral data are manipulated to produce chromatographic information and compound identification. This will be simply an overview, as LC/MS is a field in itself. But it is important for the laboratory investigator to have a working knowledge of its techniques and the future they promise.

15.1   LIQUID INTERFACING INTO THE MASS SPECTROMETER

The problem faced in the LC interface is to introduce larger volumes of mobile phase along with the compound to be analyzed into the high vacuum environment of the mass spectrometer source. LC/MS began in 1969, with a 1 µl/min flow into an EI source. The sample concentration was so much lower than the amount of solvent present that it was nearly impossible to detect the target masses of sample. Great effort was taken to increase sample concentration. Microflow systems and capillary columns were investigated as ways of increasing column efficiency, of sharpening sample bands, and of increasing concentration. These techniques were only marginally successful.

In 1970s, an LC source was developed using a continuous, moving metal band that pulled a portion of the column effluent into a heated vacuum oven and then into the mass spectrometer source for ionization. Flow rates were reduced by using a splitter in the effluent line so that the capacity of the band was not exceeded. The system worked, but volatile components were swept away with the solvent and thermally unstable compounds degraded in the drying oven.

15.2 ELECTROSPRAY AND NANO-SPRAY LC/MS

The next key advancment was the introduction of evaporative atmospheric pressure interfaces (API). Solvent is evaporated from the inlet capillary using a combination of an external heater or a nebulizing gas sleeve. Sample compounds are ionized either by chemical ion transfer or by use of a coronal discharge needle. HPLC buffers need to be replaced by volatile buffers that can be removed by the interface. The ionized compounds need to be drawn into the high vacuum mass spectrometer source through a pinhole entrance possibly protected with a curtain gas.

The atmospheric pressure interfaces for LC/MS provide real promise for general HPLC application. Originally developed for protein electrospray applications, they employ a nebulizer sleeve around the effluent inlet capillary. Injection of an inert gas, such as nitrogen, into the nebulizer provides a high velocity gas jet that breaks the atmospheric effluent into a fine mist to aid in evaporation. In the heated nebulizer interface, makeup nitrogen gas sweeps the tiny sample droplets into a electrically heated tube and then out over an ionizing coronal discharge needle. Charged sample ions are pulled by a voltage potential difference through an inert curtain gas into the evacuated source. From there the repeller forces them through the focusing lens and onto the analyzer rods.

The electrospray (ESI) and nano-spray (NSI) LC/MS system designed to work with polar and ionized compounds show tremendous application for producing multiply charged molecules. They differ primarily in the flow requirements placed on the HPLC effluent and have been applied to molecular weight determinations of protein and large peptides and show promise for analyzing DNA restriction fragments. NSI requires HPLC pump flows of $1-100$ nl/min and very fine capillary HPLC flow columns to provide maximum sample concentration. ESI is limited to microflow applications $(1-5 \mu l/min)$ in which effluent is forced through a capillary out into the source vacuum through a coronal electron discharge operating at 25,000 V. The discharge is produce off the sharp tip of a very fine needle. Electrons

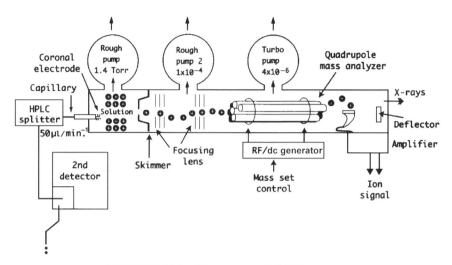

FIGURE 15.1   Electrospray LC/MS system.

released at the needle tip form a cloud through which the mobile phase stream from a capillary tube explodes into the evacuated interface. These electrons knock electrons off the sample producing molecular ions. Mobile phase buffer can cause serious plugging of the capillary tip and corrosion of the coronal discharge needle tip and should be avoided when possible in electron spray applications (Fig. 15.1).

Proteins can acquire multiple positive charges at basic amino acids such as lysine, arginine, and histidine. Since the MS analyzer separates on the basis of m/z, or mass divided by charge, mass spectrometers with a operating range of 0–2000 amu can still detect proteins with 10–50 charges per molecule.

Deconvolution of the charge envelope of multiply charged fragments developed by a single protein allows calculation of the protein's molecular weight. This can be estimated from knowledge of the fragments charges and the mass difference between adjacent fragments with incremental charges. Software is available to detect all related charge pairs through out the envelope and calculate averaged molecular weights. It is possible to detect and determine molecular weights for coeluting proteins showing overlapping fragment envelopes using this software.

15.3   ION SPRAY LC/MS

The ion spray interface (ISI) reverses the process and is designed to handle nonpolar and nonionized compounds. The nebulizer gas converts the effluent

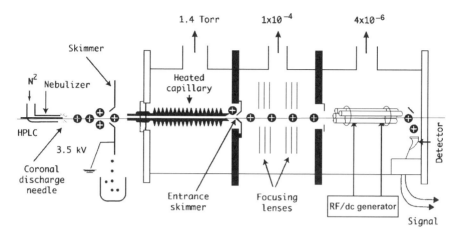

FIGURE 15.2   Ion spray LC/MS system.

into a fine mist in the presence of a high electrical potential coronal discharge. These charged small drops are swept toward a grounded liquid shield where large droplets impinge and run off. The fine, charged mist is pulled into an electrically heated capillary in a first stage vacuum chamber leading to the mass spectrometer source. After solvent evaporates, the charged molecular ions are pushed by the repeller and focused by focusing lens into the analyzer and onto the MS detector (Fig. 15.2).

This interface allows high sensitivity HPLC operation at $1-1.5$ ml/min flow rate without a stream splitter, runs gradient effluents without makeup buffer, and produces a CI molecular ion. Ion spray has the advantage of being able to produce EI type fragmentation by increasing the voltage potential between the nebulizer tip and the liquid shield, effectively creating a poor man's MS/MS system. At low voltage, mainly molecular ions are produced yielding molecular weight information. At higher voltages, fragmenting occurs to produce an EI-type pattern. A modification of this system has been published which switches voltage from high to low voltages between scans. When all scans are analyzed, fragmented data with a strong molecular ion peak are produced. In the same analysis, information is provided for molecular weight determination and for fragmentation confirmation of structure.

Dual detector LC/MS systems allow simultaneous display of UV signal and LC/MS TIC and SIC. They can be displayed with a time off set so that both types of chromatograms can be directly compared. Both can be integrated with peak detection integration using internal and external standard quantitation against calibration runs of know standards. The LC/MS data has the advantage of providing tabular peak retention and molecular weight data. Software systems are emerging that annotate the MS peaks with retention times and compound molecular weights on the chromatogram to aid in compound identification.

15.4 LC/MS/MS

Like GC/MS/MS, the HPLC interfaces can be connected to tandem and hybrid mass spectrometer analyzers to produce molecular ion fragmentation. This is of particular interest when using ISI interfaces that produce intact molecular ions. This will be of little interest and not worth the additional cost if all you only need is to identify compound by their molecular weights. However, if you need to confirm compound structure beyond that supplied by the molecular weight, you will have to fragment the parent or primary ion and run a comparison of the fragment pattern to a spectral database. Ion trap analyzers have this fragmentation capability built in and provide an economical choice for most operational laboratories. Make sure the software supplied with the LC/ITMS system is capable of running in a MS/MS mode and using spectral databases.

15.5 LC/MS VERSUS GC/MS

Liquid chromatography offers tremendous potential for analyzing nonvolatile, polarized, and ionized materials often found in real-world situations. They can separate and analyze with little or no sample purification, extraction or derivatization. Atmospheric pressure ionization interfaces offer the capability of analyzing these compounds with out having to make severe alterations of the LC chromatograph conditions to accommodate the mass spectral detector. But, it does have one major weakness: in its present form, it does not provide fragmentation data without resorting to use of a MS/MS system. It only provides molecular weights of the separated materials. Once MS detector price slides further down, an ionizing source is worked out, the library databases increase to cover nonvolatile compounds, and the tuning capability of the detector becomes automated, these systems should appear routinely on every laboratory research bench.

GC/MS has the advantage of being a more mature technology, being less expensive, and having an extensive list of established and approved operational protocols. In many areas, GC/MS is considered the "gold standard" analytical technique. In forensic work, it is the recognized, court-defensible analytical technique. For hospital and clinic analysis of therapeutic drugs and drug of abuse, GC/MS techniques are the standard that other analytical technique results must match. For environmental analysis, the US EPA has approved protocols and LC/MS results must be correlated to these standards. LC/MS protocols for specific compounds not easily analyzed by GC/MS because of stability, solubility, or size consideration are being developed, but face an uphill struggle against existing methods.

Three analytical disciplines in which LC/MS appears to be replacing GC/MS are in pharmaceutical research and manufacturing, food industry quality control, and pharmaceutical analysis hospital and clinical laboratories. All of these areas work with polarized and water-soluble compounds. The speed advantages provided by being able to analyze without extraction and derivatization strongly favor LC/MS. The field of protein and polynuclide purification, analysis, and modification are all dominated by LC/MS because of the sizes of the target molecules and their thermal sensitivity. Finally, food and agriculture analysis faces the same limitations as the pharmaceutical industry. Except for additives such as dyes, colors, and antioxidants, most of the compounds that make up our food are water soluble or complex polymeric mixture in nature. All have the best chance of being analyzed by LC/MS.

16

INNOVATION IN GC/MS

GC/MS systems are fairly well matured and are considered the "gold standard" for analysis of many compounds. The answers they provide are considered definitive proof of a compounds identity in a court of law. The most recent advances and new directions for GC/MS development are in system miniaturization and in new column development.

System miniaturization is aimed at development of portable systems that can be carried directly into the application area where they will be used. All components of the portable systems must be examined for size and weight reductions. The gas supply must be sufficient for the application, the oven replacement must fit the size of the column and be powered by a compact battery, the injector(s) must be accessible and versatile enough to supply a concentrated sample. Columns should be easily changed for flexible analysis, and the mass spectrometer must be compact, high vacuum, and capable of battery operation. The data acquisition computer must be compact, battery operated, and capable of enough processing to display an answer in the field and enough storage to return the data to the laboratory for further processing and report generation.

Column development must include surface chemistry modifications for analysis of specific compound types and also new types of support chemistry to allow ultrafast separations. Time is money in the analysis laboratory.

GC/MS: A Practical User's Guide, Second Edition. By Marvin C. McMaster
Copyright © 2008 John Wiley & Sons, Inc.

16.1   MICROFLUDICS IN GC/MS

Microfluidic control valves are beginning to find their way into GC/MS. Credit card sized integral chip-GC systems have been proposed that would contain an injector, column, interface, and column heater. A microfluidic QuickSwap interface has been marketed that connects traditional GC systems to the mass spectrometer inlet.

The QuickSwap pressure-activated control valve allows column switching without breaking the vacuum on the mass spectrometer, see Figure 6.2. This allows trimming plugged or contaminated capillary columns. More importantly to the developmental chemist, it adds ready access to another important GC control factor. In the past, the only readily accessible control factors for methods development have been carrier gas flow rate and pressure, and oven temperature ramping. Factors such as derivative formation, carrier gas selection, column selection, and packing thickness had to be determined before evacuating the mass spectrometer because of the long time required to reach an operating vacuum. With QuickSwap in place, rapid column type replacement allows separation optimization similar to that seen in HPLC. The capability of rapidly changing the chemistry of the mobile phase, the column, and compound is responsible for much of the power of HPLC. Column switching in GC has been so laborious that the tendency has been to use a standard type of column, such as a Carbowax or a phenyl bonded phase column, and play with the oven temperature ramp to achieve the best possible separation. When the mass spectrometer is added, it is often used to try and resolve the identity of overlapping, unresolved peaks. Changing the chemistry of the column can often provide a much better resolution and simplify the job of the mass spectrometer.

So far, QuickSwap is designed to be actuated by an electronic pressure control (EPC) system, which limits it to a single manufacturer's systems. If it becomes as useful as it appears at the moment, it should be possible to retrofit it to other GC/MS systems.

The chip-GC is much more a concept than a product such as QuickSwap. An integrated silica-based chip-GC appears to have first been proposed in a 1975 Sanford University Ph.D. dissertation by SC Terry. Micro-GC development seems to be aimed at analysis of chemical warfare agents, hazardous chemicals detection, alcohol breathe analyzers, and extremely low weight components for space exploration. A 2005 paper presented at the AICHE annual meeting reported that the University of Illinois, Department of Chemical and Bioengineering, has created an on-chip micro-GC with a carbon nanotube sensor for detecting chemical warfare agents. This credit card sized GC could fit into an interface box mounted on a high sensitivity mass spectrometer, probably a tiny quadrupole unit or, more likely, an ion trap. The injector would probably have to be a headspace analyzer/concentrator and the column a packed or wall-coated etched channel such as used in commercial chip-LC modules. A web site to watch for further development of this type of product is www.gcms.de.

16.2   RESISTANCE COLUMN HEATING

Traditional GC systems are built around the GC oven containing the oven, the injector, the column, and the system controller. The detector may be mounted on top of the oven or made to stand beside it. The oven and its heater are the largest component of the GC and the obvious first target for size and energy reduction. The first step is a hollow, hinged pair of electrically heated metal plates that surround the coiled column. Resistance heating leads to a problem of heat dissipation that is usually controlled by adding a circulating fan, which also helps in GC re-equilibration at the end of a chromatography run. An alternative would to use Pelltier-type of heating/cooling of the column plates, as used in modern detector temperature control, which might reduce the overall energy demand on the battery system.

Portable systems would need a syringe injection port, either a septum type or one of the nonseptum valve injection ports. The field GC/MS has often been called the substitute for the drug-and bomb-sniffing dog. As such, it will need some type of headspace injector to sample and concentrate gasses in the environment for analysis.

16.3   PORTABLE GAS SUPPLY

Carrier gas supply will initially be from a high pressure mini-lecture bottle and valve source. Advances under consideration for hydrogen-fueled vehicle energy sources may offer a form of high density hydrogen storage that could be used to provide GC/MS carrier gas. Metal hydride and ceramic clathrate foams offer promise to allow safe storage of large quantities of hydrogen and controlled gas release by low energy heating of the casing around the foam.

16.4   PORTABLE GC/MS SYSTEMS

Ten years ago I read a discussion of a quadrupole MS unit under development using 5-in. long hyperbolic glass or sapphire rods. Viking 573 or SpectraTrak is based on a portable Agilent quadrupole mass spectrometer. Hapsite is a field-portable GC/MS from Indicon, Inc. weighing 37-lb and can be worn over the shoulder to create a man-portable GC/MS. It uses a pumped sampling wand and can be operated for about 10 min to analyze a wide list of compounds that can be analyzed with a NIST spectral database.

The www.gcms.de. web site lists a number of microsystems based on other types of mass spectrometers. Griffin Analytical Technologies has a breath analyzer based on a cylindrical ion trap, as is the Chemical Biological Mass Spectometer (CBMS) from Bruker. Kore offers a fire scene investigation system called MS-200 based on a time-of-flight mass spectrometer.

Portable GC/MS systems come in a variety of sizes. They start out as transportable units that fit in the back of a pickup and are powered by the truck's 12-V battery system. Luggable systems, like the Hapsite unit, are versatile systems designed for general Hazmat and first response analysis of toxic spills and suspected chemical warfare sites. Smaller handheld units generally are targeted at very specific analysis, such as bomb and explosive sniffers or breath alcohol and drug analyzers.

16.5   NEW COLUMN TECHNOLOGY

Monolith and micro-WCOT (wall-coated open tubule) columns are the bleeding edge of new GC column technology. Monoliths and micro-WCOTs are not packed columns, but are prepared *in situ* in open tubule and capillary column casings. In monoliths, the interior of the capillary tube is completely filled with either a polymeric or silica foam with interconnecting pores. The porous foam structure can be coated with a viscous liquid phase or permanently reacted with a bonded liquid phase. Micro-WCOT column eliminates the solid support structure and reacts the organic-bonded phase directly to the walls of the column to prepare a brush phase directly adhering to the wall of a ultrasmall diameter column.

Organic polymer monoliths were first prepared in the 1990s by poly-merizing a cross-linked monomer in the presence of a poromeric liquid. After polymerization catalyzed either chemically or with UV light, unreacted monomer, catalysts, and the poromeric liquid are washed out with an organic solvent such as tetrahydrofuran. Functionality can be added to monomeric column by selecting monomers with the desired side chains. Because the foam completely fills the column, diffusion effects due to void volumes are avoided leading to much high column efficiency.

Silica monolith columns began to appear early in 2002. Tetraalkyoxysilane is polycondensed in the presence of porogenic liquid such as poly(ethylene glycol). After washing out the porogen, the monolith is left behind as a stiff rod with connecting through pores and internal mesopore that provide a very large surface. Considerable shrinkage occurs in preparing the fragile 4.6 mm × 15 cm columns that must be covered with a PEEK polymeric sleeve and forced into a supporting metal column. Monoliths can be prepared directly in capillary tubes, washed out, reacted directly with an organic-bonded phase to provide columns efficiencies similar to that shown by very small diameter bonded-phase packed columns, without the ultrahigh back-pressures shown by 1.7 μm silica beads.

High efficiency WCOT columns are still in early stages of development. Large-diameter capillaries WCOTs have been around since the 1970s with the bonded organic phase substituting for a coated phase and provide high efficiency gas analysis. Microcapillary column internal diameters must be in the range of 20–50 μ, filled with bonded organic brush phases such as octyldecyl, usually through a C_c to C_{12} alkyl spacer. The capillaries would be supported in a rigid casing bundled with many other capillaries, similar in design to osmotic water filtration tubes. Design of common micro-inlet and micro-outlet throats for the capillary bundle must be completed in order for these columns to be mated to injectors and detector for comparison to the efficiencies shown by silica and polymeric monolith capillaries.

APPENDIX A

GC/MS FREQUENTLY ASKED QUESTIONS

I have made a list of common GC/MS questions I hear from students and customers and the answers that I found. This list is not exhaustive. I have tried to leave out some of the more inane questions that I have received. One of the most common questions that I did not include was: "Why won't my system startup?" *Treatment:* I would ask the person on the phone if the system was plugged in. After the explosion on the other end settled down, I would say, "Sir, sometimes the janitors unplug lines so they can plug in their polishers. Would you please check to see if it is plugged in?" Usually after about a minute or so I would hear a quiet click as the phone was hung up.

A.1   GC FAQs

1. *Why use helium gas instead of hydrogen? Doesn't hydrogen usually provide sharper peaks?* Hydrogen is explosive. Most laboratory safety officers are members of the Hindenburg society, realize how poorly ventilated most laboratories are, and will not permit use of hydrogen.

2. *Where is the injector on my headspace analyzer?* The headspace sampler is the injector. It has an inlet and an exhaust valve for filling the sampler reservoir and a carrier gas port for purging the reservoir onto

GC/MS: A Practical User's Guide, Second Edition. By Marvin C. McMaster
Copyright © 2008 John Wiley & Sons, Inc.

the column. Systems not dedicated totally to headspace analysis will often have a secondary injector plumbed into the exhaust line before the column head.

3. *How do I use the injector to automatically start my data system?* There is a sensor ring option for most injectors that fits around the injector port. When an injection is made, the ring is pushed down making a contact closure that sends a signal to the data start port on your integrator or computer system.

4. *My injector seal seems to be leaking sample. How often should it be replaced?* Before you start losing sample! Recommendations are every 100 injections or every 2–3 days (see chapter 9).

5. *Do I need an autosampler?* It depends on your sample load. Laboratories that run samples on three-shift 24-h days almost always use autosamplers. University laboratories with lower system use prefer graduate students. It is easy to justify the price of the autosampler in cost per test operation.

6. *Equilibrating my oven takes as much time as my run. Can I shorten this time?* The easiest way is to open the oven door at the end of the run. Automated version of this technique pushes the door open with a piston until your oven temperature drops below you initial temperature, then pops it shut with a spring. Cryoblasting systems use adiabatic cooling using carbon dioxide gas purging. Your oven usually has to be purchased with either of these last two options. Opening the oven door is available for all systems.

7. *Why does my GC oven controller have a series of curves that can be selected for programming?* Early oven controller had limited micro-processor memory. The preprogrammed curves let you simulate complex multislope gradient profiles. A curve that rose rapidly and then converted to a plateau would make early running peaks come off more rapidly than late runner compared to a completely linear temperature slope. A curve that had an early plateau followed by a sharp final rise will compact late running peaks. A few manufacturers have retained the curves as programming segments so that truly complex gradients can be created.

8. *Will electrically controlled heating/cooling blocks replace ovens?* They certainly take up a lot less space, but for the moment they restrict the size and type of GC columns you can use and they are more expensive. Peltier-type temperature controllers are usually sized to enclose fairly small capillary or monolith columns. Their major applications have been in microfluidic, man-portable, and extraplanetary GC/MS systems where price is not as great an objection.

A.2   COLUMN FAQs

1. *Which type of column should I use in my methods development?* Obviously, the liquid phase depends on what you are trying to separate. Protocols usually specify the type of the column and its working phase. Since you specified methods development, I would select a bonded-phase capillary column. Bonded-phase columns have a longer lifetime with less column bleed into the MS detector. Bonded-phase monolith columns are touted as the column of the future, but they still need development work. Capillary columns can be shortened for cleaning the column head and coiled and squeezed to fit temperature-controlled column heaters.

2. *Why can't I use column switching as a control variable in GC like I can in HPLC?* Column switching in conventional GC systems requires disconnecting the interface to the mass spectrometer and breaking the vacuum. Vacuum restoration can take up to 4 h. Aligent has announced a microfluidic device called QuickSwap to divert carrier gas flow away from the mass spectrometer allowing the vacuum to be maintained while the column is replaced. This is controlled by their EPC system, but if it can be adapted it may offer a general method for rapid column switching in GC/MS methods development.

A.3   MS FAQs

1. *Which mass spectrometer is best; quadrupole, ion trap, or time-of-flight?* Each analyzer has its own strength. Quadrupoles were the first mass spectrometers used for chromatography, still represent the bulk of laboratories systems, and are the least expensive. Look for a bench-top quadrupole that can give you chromatograms labeled with peak molecular weights. Linear ion trap analyzers are the most flexible allowing you to do simple molecular weights as well as precursor and product fragmentations studies. They also can store enough ions to give excellent sensitivity for trace analysis and can trap specific ions for fragmentation analysis. For large molecule studies you are going to want a time-of-flight system with an electrospray interface.

2. *How high of a vacuum should I be getting?* You will need at least 10^{-5} Torr before you begin running chromatography and should see 10^{-6} Torr on a clean system with a turbo pump. I have talked to people who get 10^{-7} Torr with an oil diffusion pump and clean oil. Sometimes you need to push down on the lid on the mass spectrometer containment

system if you are having trouble making a vacuum after cleaning the system. Try changing the pump oil in your rotary-vane pump if you are not able to reach 10^{-5} Torr with the turbo pump.

3. *How do I service my systems turbo pump?* A turbo pump is a miniature jet engine and should only be worked on by a trained technician. The best plan is to pay for an exchange program warrantee with the system manufacturer. When a turbo pump is down, it is down. It is nothing you want to mess with repairing.

4. *How often do I need to replace my ion detector?* Cascade-type detectors have a finite lifetime and need to be replaced periodically. Record the EM voltage needed to tune a new detector and watch that value as you continue to autotune. When it exceeds 3000 V, look for a replacement. Time-array detectors used on TOF systems have fewer problems, but are more expensive. Problems show up as increased background noise due to increasing element chatter.

5. *How do I know when I need to clean my mass spectrometer?* Dirty systems lose sensitivity and become very difficult to autotune. The source filament and lens are the major cleaning problems in a GC/MS where most of the problem is char due to filament-based sample burn. The system also accumulates organic contamination on the analyzer and detector surfaces. Laboratories that I have been in disassembled a quadrupole system, immersed the analyzer rods still in their ceramic collars in a graduated cylinder, and flooded them with methanol. They wiped the detector face with a lint-free paper wetted with methanol. They dry them in an evacuated vacuum desiccators if the rods are small enough. If not, they wipe them with lint-free paper and dried them by pumping overnight in the mass spectrometer containment chamber.

6. *All I want is molecular weight chromatograms? If I am running in the CI mode will my peaks be labeled?* The peaks will be labeled with molecular weights if you select the correct software. The CI mode is low energy ionization sources and generally does not fragment the molecular ion. Some software will select the major fragment ion and label the TIC with major ion molecular weights if you are running a quadrupole or an ion trap detector in the CI mode with carbon dioxide in the ring electrode.

7. *How come I don't see a molecular ion peak in my chromatography?* A mass spectrometer running in the EI mode uses 70-eV electrons to ionize molecules in the carrier gas from the GC or the jet separator. The high energy molecular ions formed are generally unstable and break down to form the fingerprint fragmentation pattern that is used to

identify the structures of the compound when we send it to a library search. Generally, if the molecular ion is there, it is present as only a very tiny peak.

A.4   GC/MS FAQs

1. *How do I start up my system after it has been sitting unused for a while?* Check the operator's manual that came with the system. I would start up a system I am unfamiliar with using the following procedure. It really depends on whether the vacuum system on the mass spectrometer has been turned off. If it has, you will need to connect the column, start the rotary-vane oil pump until you reach a vacuum of 10^{-4} Torr and then turn on the turbo pump and begin to establish your analyzer vacuum. This may take a number of hours.
   Check to see if the GC column is in place. If not, select the column designated by your procedure and place it in the oven. Connect the column to the injector and the splitter inlet before the mass spectrometer. Turn on the carrier gas, select your flow rate, and set the oven program for your chromatography run from the oven controller.
   When your vacuum has reached at least 10^{-5} Torr. Turn on the splitter gas, open the calibration gas valve, and run an autotune on the mass spectrometer. By now your column should be equilibrated and you are ready to make an injection of a calibration standard. Set your mass range selection for scanning and make your injection or start your autosampler and data acquisition from the control computer.

2. *Do I need a data library to identify my compounds?* Spectral data libraries, such as the NIST or Wiley libraries, are needed to help identify unknown compounds from fragmentation data. If you are an expert on spectral analysis, you might consider bypassing use of a library.
   If you are running the same compounds each day, you may need to only look for known fragment patterns. If you are running environmental assay, you must have a library database to search for unknown compound to generate a tentatively identified compound report.

3. *Can I fix up my GC/MS to do LC/MS?* The first LC/MS systems were made from a mass spectrometer converted from a GC/MS, so you should have little problem. You will need an HPLC system and an ionization interface, so this would not be a trivial conversion. It really depends on how much you use the GC/MS system. Switching back and forth will rapidly become a pain. At a minimum, you will have to shut

off your vacuum, remove the GC column, insert the ionizing interface probe, and then reestablish the mass spectrometer vacuum. You might want to consider writing another grant proposal and buying a dedicated LC/MS system after you have made the conversion one or two times.

4. *What is the difference between calibration and autotuning?* Calibration involves adjusting the mass axis to align it with known peaks positions from an injected standard using tuning lens, amu off-set lens, and the electron voltage of the detector. Tuning involves balancing the settings on the same lens to adjust the relative heights of the calibration peaks to a predetermine relationship. Autotuning is done by the system controller adjusting the lens, among others to produce a specific calibration and tune when calibration standard is injected through the interface port or from the LC system. The operator's manual for your particular system should recommend calibration standards and tuning criteria. If it is not in the manual, contact their technical support or their web site. Your instrument's calibration and tuning need to be checked periodically if your MS results are to be compatible with those from other instruments and useful for fragmentation database library searches.

5. *When do I need GC/MS/MS and when will GC/MS be good enough?* A GC/MS system will provide you with total-ion chromatograms of your column peaks and their solvent complexes. Fragmentation patterns can be extracted from the peaks and referred to a spectral database for compound identification and compound molecular weights. A GC/MS/MS systems will be much more expensive, but it can provide structural information of daughter ions of the original fragmentation to confirm the identity of the separated compounds, trace materials, and breakdown mechanisms. But, it will do it at a cost. Someone will have to plan the fragmentation study and interpret the results. Library database may be able to make this identification for you for simple compounds, but generally each sample run will be something of a research project.

6. *Can I use my NIST library to identify compounds in my chromatogram?* A GC/MS system generally fragments the sample in the carrier gas but usually does not provide a single ion equivalent to the molecular weight. Fragmentation information can be searched by your NIST library to provide best fits from the database to provide the compounds structure and molecular weight if the compound is in the library. You will have to examine the possibilities offered by the library software to see if they make sense. It is important to remember that the NIST, Wiley,

and other commercially available libraries were originally developed from runs made on magnetic sector and quadrupole GC/MS systems. The best check on the library is to run the compound that provides the best fit on your GC/MS under your specific operating conditions to see if you get fragmentation data that match your original run.

APPENDIX B

GC/MS TROUBLESHOOTING QUICK REFERENCE

This section is designed to assist in troubleshooting system problems. It is not meant to replace systematic troubleshooting and routine maintenance. A systematic reverse order approach to problems is always better. Keeping this in mind, the commonly seen problems, possible causes, and suggested treatment are listed in this appendix. Appendix C contains a list of common background contaminants.

B.1 GC INJECTOR PROBLEMS

Problem 1: Peaks broaden and tail.
    Cause a: Poor column installation causing dead volume in the injector.
        Treatment: Reinstall column in injector. Check seal at ferrule. Check insertion depth. Ensure a good column cut.
    Cause b: Solvent flashing in hot injector.
        Treatment: Reduce injection speed on hot injectors and if possible reduce injector temperature. If you are using sandwich injection, reduce solvent plug to 0.5 µl.
    Cause c: Incorrect injector temperature control.
        Treatment: Typically set injector temperature at 20 °C lower than solvent boiling point and keep column at solvent

GC/MS: A Practical User's Guide, Second Edition. By Marvin C. McMaster
Copyright © 2008 John Wiley & Sons, Inc.

boiling point. Hold column at initial temperature until injector has finished heating.

Cause d: Septum purge line is plugged.
Treatment: Check that the septum purge flow is ~0.5 ml/min. Change the septum purge frit or adjust the needle valve.

Cause e: Injector not being purged properly after splitless injection.
Treatment: For splitless injection, the vent flow should be 70 ml/ min, and the injector should be switched to the split mode 0.5–1.5 min after injection.

Problem 2: Tailing sample peaks for active components.

Cause a: Active sites in the injector insert or liner.
Treatment: Change or clean the injector insert. Silanize it, if necessary.

Cause b: Active sites or degraded phase in column.
Treatment: Remove the front 15 cm of the column and reinstall. If retention times are changing or cutting the column does not help, replace the column.

Cause c: Injector not hot enough for higher boiling compounds.
Treatment: Increase the injector temperature and lower the injection speed. Check that the graphite ferrule is free of cracks and the septum support is tight.

Problem 3: Low response and tailing of high boiling point compounds.

Cause a: Injector is not hot enough to vaporize high boilers.
Treatment: Increase injector temperature.

Cause b: High levels of column bleed are masking component peaks.
Treatment: Condition column or change to a high temperature column if conditioning does not help. Consider changing to a bonded phase column if problem continues.

Cause c: High levels of silicone is coated on ion source surfaces.
Treatment: Clean the ion source.

Cause d: Interface/ion source not getting to adequate temperature.
Treatment: Change the manifold heater.

B.2 GC COLUMN PROBLEMS

Problem 4: Leading sample peaks.

Cause a: Column overload due to excess amount of component injected.

        Treatment: Dilute the sample or do split injection.
    Cause b: Degradation of stationary phase.
        Treatment: Change the column. Change to a bonded phase column.
    Cause c: Carrier gas velocity too low.
        Treatment: Increase carrier gas flow rate.

Problem 5: Poor chromatographic resolution.
    Cause a: Column temperature or program not optimized.
        Treatment: Modify method by changing temperature ramp segment slopes. (See GC methods development in chapter 6.)
    Cause b: Carrier gas flow rate not optimized.
        Treatment: Decrease carrier gas linear velocity.
    Cause c: Column not capable of the separation.
        Treatment: Change to a more polar column. Change to a capillary column with a higher plate count.
    Cause d: Stationary phase has degraded.
        Treatment: Replace the column.

Problem 6: Peak size changes from run to run.
    Cause a: Leaking or partially plugged syringe.
        Treatment: Check visually that the syringe is pulling up sample. Remake Teflon seal around the autosampler syringe needle or flush the syringe with solvent. Heating the syringe in a hot injector may help if it is plugged; otherwise, replace the syringe.
    Cause b: The septum leaks.
        Treatment: Replace septum regularly. Insure that the septum nut is tight.
    Cause c: Improper column installation in injector or column inlet leak.
        Treatment: Check installation of column in injector and tighten the capillary column nut.
    Cause d: Sample is absorbed by active surfaces in injector or column.
        Treatment: Change injector insert and deactivate it if necessary. Remove front 5 cm of column, if it is a capillary column or replace the column.
    Cause e: Sample is incompletely vaporized in the injector.
        Treatment: Increase the injector temperature or the maximum programmed temperature of the injector.

Problem 7: Peak splitting, especially low boilers.

Cause a: Sample is flashing in the injector simulating two injections.
          Treatment: Lower injector temperature. Use sandwich technique
                for splitless injection.
Cause b: Column temperature program starts before injector is heated.
          Treatment: Increase initial column hold-time until injector has
                reached its maximum temperature.
Cause c: Solvent plug in injector.
          Treatment: Decrease solvent plug to 0.5 µl or eliminate if
                possible.

Problem 8: Extra peaks in chromatogram.
Cause a: Septum bleed, particularly during temperature programming.
          Treatment: Use high temperature, low bleed septum. Ensure that
                septum purge flow is 0.5 ml/min.
Cause b: Impurities from sample vials such as plasticizers.
          Treatment: Confirm by running solvent blank with new syringe.
                Change to certified sample vials and keep samples
                refrigerated.
Cause c: Impurities from carrier gas.
          Treatment: Install or replace carrier gas filters.
Cause d: Contaminated injector or GC pneumatics.
          Treatment: Remove column from injector and bake out at elevated
                temperature with a purge flow of at least 20 ml/min.
Cause e: Impurities in sample.
          Treatment: Confirm by running a blank or standard run.

Problem 9: Retention times shift in chromatogram.
Cause a: Unstable carrier gas flow controller/regulator.
          Treatment: Check pneumatics for leaks. Replace flow controller/
                regulator if necessary.
Cause b: Column contamination or degradation.
          Treatment: Condition or replace column.
Cause c: Leaks at septum or column to injector connection.
          Treatment: Replace septum regularly and check that the septum
                nut and the capillary column nut are tight.

B.3 MS VACUUM AND POWER PROBLEMS

Problem 10: High vacuum pump would not turn on. No pressure reading.
Cause a: Filament is burned out. No filament voltage.
          Treatment: Shutdown vacuum. Replace filament.
Cause b: System leak. Rough pump cannot reach starting vacuum.

Treatment: Find and repair system leak.

Cause c: Rough pump would not turn on.

Treatment: Check rough pump power. Replace pump oil. Replace pump.

Cause d: Turbo pump rotor has seized.

Treatment: Turn in turbo pump for replacement.

Cause e: High vacuum gauge cathode burns out.

Treatment: Replace cathode tube.

Problem 11: Cannot reach operating vacuum (10^{-6} Torr).

Cause a: Contaminated fore or diffusion pump oil.

Treatment: Look for background increase in previous TIC. Replace fore pump oil. Have diffusion pump oil replaced by service representative.

Cause b: Analyzer contaminated by diffusion pump oil.

Treatment: Shut down mass spec. Disassemble. Clean quadrupole rods with methylene chloride, acetone, or methanol. Buy a mass spectrometer with functional butterfly valves.

Cause c: Major air leak around column fitting into interface.

Treatment: Replace column ferrule and reseat compression fitting.

Cause d: O-ring around analyzer housing is not seating.

Treatment: Push down on analyzer housing cover with vacuum on. There should be a change in rough pump sound as vacuum increases. Replace housing gasket.

B.4   MS SOURCE AND CALIBRATION PROBLEMS

Problem 12: With repeller at maximum, cal gas 502 cannot be found.

Cause a: Source is dirty.

Treatment: Shut down vacuum. Clean ion source and lens.

Cause b: Mass axis is badly out of calibration.

Treatment: Autotune. If 502 is still off scale, recalibrate mass 131 against 69, then 264 to 131. Cal gas 502 should be on scale. Calibrate 502. Recheck the 131 and 69 masses.

Cause c: Detector showing loss of sensitivity/burn out.

Treatment: Increase EM voltage to 3500 V. If still no 502, replace the detector.

Problem 13: Calibration gas' 69 peak position moving. Poor 70 mass resolution.

    Cause a: System grounding problem.

        Treatment: Ground analyzer unit to control interface. Float the computer ground.

    Cause b: Ion source is dirty.

        Treatment: Shut down system. Clean the source.

Problem 14: No calibration gas peaks.

    Cause a: Cal gas valve not open.

        Treatment: Open cal gas valve.

    Cause b: Calibration gas solenoid valve stuck open. All calibration gas evaporated.

        Treatment: Have solenoid replaced. Put fresh PFBTA in the cal gas vial.

B.5 MS SENSITIVITY AND DETECTOR PROBLEMS

Problem 15: Analysis sensitivity has decreased.

    Cause a: Background has increased.

        Treatment: Check column bleed, septum bleed, pump oil, and ion source contamination.

    Cause b: Detector needs replacement.

        Treatment: EM voltage is over 3500 V to see cal gas 502. Replace detector.

APPENDIX C

SOURCES OF GC/MS BACKGROUND CONTAMINATION

The GC/MS is an extremely sensitive instrument. However, the achievement of this kind of sensitivity is background dependent and requires elimination of all common sources of contamination. Essentially, two kinds of backgrounds can interfere with trace-level GC/MS analyses:

1. *General background contamination,* such as column bleed, hydro-carbons, and phthalate plasticizers, which will generate a large TIC signal during the analytical scan and decrease the sensitivity level for detecting target compounds.
2. *Specific ions in the background* will interfere with a single-ion or extracted ion chromatogram. For example, significant 164 background might be present when trying to detect low levels of 2,4-dichlorophenol. This type of problem is less common than general background con-tamination. Typically, a single ion or an extracted ion can be chosen which does not appear in this background.

The easiest way to determine if the background is permanent, is to lower GC temperatures to 50 °C and run a scan to see if the background decreases. If it does, the background is probably due to column bleed, septum bleed, contaminated pump oil, or leaks of various kinds.

GC/MS: A Practical User's Guide, Second Edition. By Marvin C. McMaster
Copyright © 2008 John Wiley & Sons, Inc.

In all instances where the background is determined to be coming from the analyzer and not eluting from the GC, the system should be shut down and the source cleaned. If this does not eliminate the problem, shut down the system and dip the rods, washing with methanol or methylene chloride to remove contaminants. A permanent background is defined as background that is at approximately the same level, regardless of GC temperatures.

The other source of problems contaminated samples tend to give move discrete chromatographic peaks and specific mass fragments. These samples can be cleaned or removed with SFE or SPE cartridge or GPC columns before injection.

The GPC columns separate on size and release smaller molecules before the larger, polymeric material. They are very good for removing "road tar like" materials from your extracted samples. Although getting the road tar off the column may prove to be a problem, generally, if it can be dissolved, it can be eluted.

The SFE or SPE cartridge columns are true chromatography columns. They can be used to do class separations of materials. Using windowing techniques and standards, you can work out methods for purifying the materials of interest from either polar or nonpolar contaminates. This technique is described in "HPLC: A Practical User's Guide" (see Appendix E). Finally, their cost is low, and if contaminated and the contaminant cannot be removed, they can be discarded.

Following is a list of some common contaminant mass ion:

Mass ions	Compounds	Source of origin
18, 28, 32, 44	H_2O, N_2, O_2, CO_2	Air Leak
28, 44	CO, CO_2	Hydrocarbon fragments
31	Methanol	Lens-cleaning solvent
43, 58	Acetone	Cleaning solvent
69	Fore pump fluid	Saturated trap pellets
69, 131, 219, 254, 414, 502	FC43 (PFTBA)	Calibration gas leak
73, 207, 281, 327	Polysiloxanes	Column bleed
73, 207, 281, 149	Polysiloxanes	Septum bleed
73,147,207,221,295,355,429	Dimethylpolysiloxane	Septum breakdown
77	Benzene or xylene	Cleaning solvent
77, 94, 115, 141, 168, 170, 262, 354, 446	Diffusion pump oil	Improper shut down of pump heater
91, 92	Toluene or xylene	Cleaning solvent
105, 106 xylene	Cleaning solvent	
151, 153 trichloroethane	Cleaning solvent	
149	Plasticizer (phthalates)	Vacuum seals damage
14 amu spaced peaks	Hydrocarbons	Saturated trap pellets, fingerprints, pump fluid

APPENDIX D

A GLOSSARY OF GC/MS TERMS

Base peak   The most intense ion fragment in a compound's spectrum under a given set of experimental conditions.

Capillary zone electrophoresis (CZE)   A separation technique based on movement of ionized compounds through a capillary tube filled with buffer toward a high voltage of the opposite polarity. Separation is based on the compound's size and charge potential.

Carrier gas   Gas used to sweep volatile materials from the injector, through the GC column, and on into the detector.

Chemically induced (CI) ionization   Ionization in a MS source in which a diluting gas, such as carbon dioxide, is added to the analysis sample. The diluting gas, being in higher concentration, is ionized first and transfers this ionization to the sample at a low energy forming a more stable molecular ion. Used in molecular weight determination.

Chip-GC   A microfluidic device having all the components of a gas chromatography on an integrated circuit.

Column   A packed tube filled with coated, absorptive stationary phase particles used to achieve GC separations.

Data/control system   The "brains" of the GC/MS system which programs the system components, controls MS scanning and lens, and acquires and processes the data from the detector.

GC/MS: A Practical User's Guide, Second Edition. By Marvin C. McMaster
Copyright © 2008 John Wiley & Sons, Inc.

Detector   A device that produces a voltage change in response to a change of composition of the material in its flow cell.

Differential pumping   An arrangement in which two chambers connected by a small orifice, like an MS source and analyzer, have two pump connections through different diameter exhaust tubes. Capable of providing different pumping rates and vacuums in the two chambers.

Direct insertion probe (DIP)   A metal probe with a slanting flat surface which allows sample to be inserted through a vacuum port directly into the ionizing electron beam in the MS ion source.

Efficiency factor   A chromatography resolution factor that measures the sharpness of peaks. Control by carrier gas nature and flow rate, particle size, coating thickness, and column diameter and length.

Electrode   A source of electrons for ionizing samples. See also filament and ring electrode.

Electron induced (EI) ionization   Sample ionization in an MS source by bombardment with 70-eV electrons from a filament. This is a high energy ionization leading to fragmentation of the original molecular ion.

Fast atom bombardment (FAB)   Ionization for nonvolatile samples. Suspended in glycerol, the sample is placed on a DIP probe tip and inserted into a stream of heavy metal ions in the source. The matrix explodes vaporizing the ionized sample, which is repelled into the analyzer.

Filament   Metal plates that connect to the ion source and release a stream of ionizing electrons when a voltage charge differential is applied.

Fourier transformed GC/MS (GC/FT-MS)   A separation technique in which a GC sample is ionized in an evacuated chamber, held in place by a cyclonic trapping voltage, excited to a higher orbit by a "chirping" multifrequency signal, and transmits an RF signal characteristic of all the masses present. Transformation of this multifrequency signal allows plotting of intensity versus m/z spectra with very high sensitivity at each chromatographic point.

Gas chromatography (GC)   Separation technique in which the volatile analyte is swept by a carrier gas down a column packed with packing material coated with an absorbing liquid. Differential partition between the two phases by sample components leads to band separation and elution into a detector.

Injector   A devise used to move a sample in an undiluted form onto the head of a column.

Internal standard   A compound added during the last dilution before sample injection in equal concentration to all analyzed samples. Its purpose is to correct for variations in sample injection size. It also can be used to correct for variations in peak retention times.

Ion trap detector (ITD)   A desktop MS that ionizes and holds the ionized sample with in a circular electromagnet until swept with a dc/RF frequency signal that releases the ionized sample into the ion detector.

m/z   A symbol for mass divided by charge, measured in amu or daltons. The *x*-axis for a mass spectrum indicating that a MS spectra is dependent both on the mass and the charge on the fragment ion.

Matrix blank   A quality control, matrix-only sample analyzed to show all levels of target compound present before spiking with standards.

Matrix spike   An environmental analysis QC sample required for 5% of all sample analyzed. A matrix blank is spiked with all standards at a level within the analysis range and checked for recovery of standards.

Molecular weight   Summation of the weights of all the elements in a molecule expressed in amu or daltons. In MS, the *m/z* value of the molecular ion in EI ionization.

MS/MS GC/MS system   Most commonly, an ion trap or tandem, triple-quadrupole system for study of MS fragmentation mechanisms. In a tandem system, the second analyzer is a collision cell used to further fragment ions separated in the first analyzer for analysis in the third analyzer. Ion traps have inherent MS/MS natures. Hybrid MS/MS systems are made up of marriages of a variety of mass spectrometer types.

On-column injection   GC injection directly on to the head of the column used to avoid loss of sample in the injector due to thermal breakdown. Has problem of contamination of column by nonvolatile sample components.

Pascal (Pa)   A measure of pressure equal to 7.5×10^{-3} Torr (mmHg). Pressure measurement commonly used in Europe, but also used by some US manufacturers.

Quadrupole analyser   Mass spectrometer analyzer based on four cylindrical rods held in a hyperbolic configuration and swept with a variable frequency dc/RF signal allowing selection of individual mass fragments.

Qualifiers   Other major fragment peaks in a compounds spectra; with the target mass, used to confirm the identity of the compound. The qualifier's mass and its height relative to the target mass are used to confirm identity.

QuickSwap   A microfluidic pressure-controlled switch allowing column changing while maintaining the mass spectrometer's vacuum.

Reagent blank   A first blank run to indicate the cleanliness and the capability of laboratory to run samples. Reagent water is spiked with all standards and subjected to full analysis conditions.

Resolution equation   A measure of a column's separating power. It combines retention, separation, and efficiency factors into a single equations that shows their interactions.

Retention factor   A column resolution factor measuring how separation is effected by residence time on the column. Controlled by temperature and carrier gas pressure.

Retention time   The length of time a compound stays on the column under a given set of experimental conditions.

Ring electrode   The central electrode of an ion trap used to hold ion fragments in circular orbits until the time to elute them into the detector.

Roughing pump   The first pump in a vacuum system. It is used to reduce pressure initially from atmospheric pressure to a low pressure that can serve as a starting point for the high pressure pump; usually a mechanical rotary-vane pump.

SCAN   An MS operational mode in which the amount of each mass unit is measured by continuously changing the dc/RF frequency on the quadrupole. Mass can be scanned low to high or high to low. The latter leads to less intermass tailing and more accurate relative height measurements.

Separation factor   A column resolution factor control by the column's chemistry and by temperature. Changes in this factor result in shifting of relative peak positions.

SIC   Single-ion chromatogram. Chromatogram produced by displaying the ion current produced versus time for a given mass (m/z). It can be produced by operating in a single-ion mode or extracted out of scanned fragment database.

SIM   Single-ion monitoring. The mass spectrometer measures one or a few specific masses. Since fewer measurements are made than in SCAN mode, they are made more often with a proportional increase in sensitivity.

Spectra   A plot of signal intensity, measured in volts, versus ion fragment m/z, measured in amu, for a given MS scan or range of scans. The data are usually summed around unit mass and presented as a bar graph of intensities relative to the base peak.

Supercritical fluid chromatography (SFC) A column separation technique using pressure/temperature control to convert a gas into a fluid that is used as the mobile phase for liquid/solid chromatography. Sample recovery is achieved by releasing the pressure to turn the mobile phase back into a gas releasing the dissolved sample.

Surrogate A standard compound added in known amounts to all processed samples. Its purpose is to detect and correct for sample loss due to extractions and handling errors. Usually it is a deuterated or other labeled chromatographic equivalent of an analyzed compound, not normally found in nature.

Target compound quantitation Quantitation based on identifying a compound by locating its target and qualifier ion fragments. Once identified, the target ion signal strength is compared to known amounts of standards to determine the amount present.

Target ion A compound's MS ion fragments chosen for identifying and quantitating the amounts of the compound present in mixtures of standards and unknowns. Usually, but not always, the major fragment ion in a compound's spectra.

Temperature ramp A gradual, controlled increase of temperature with time. It is used in combination with holds and other ramps in building a oven temperature program for resolving compounds on a GC column.

TIC (1) *Total ion chromatogram.* A chromatogram produced by measuring the total ion current from the mass spectrometer versus time. A **TIC** data point represents a summation of all mass fragments present at a given time.

     (2) *Tentatively identified compound.* A compound, found in the chromatogram of an unknown, not listed as a target compound, internal standard, surrogate, or known compound. It is referred to library search and a reasonable number of matching compounds are reported.

Time-of-flight GC/MS (GC TOF/MS) Chromatographic technique in which the MS detector analyzes effluent mixed with a chromaphor is burst-ionized with pulsed laser energy bombardment and components mass fragments are identifyed by the time they take to travel a flight tube and reach a detector. LC/TOFMS is becoming popular in the analysis of charged biochemicals, proteins, and DNA restriction fragments with multiple charges.

Torr   A commonly used measure of pressure of vacuum equal to 1 mmHg or 133.32 Pa.

Triple-quadrupole GC/MS/MS   A tandem quadrupole system in which a gas chromatograph feeds a mass detector with three quadrupole units in series. The second quadrupole acts as a holding and collision cell in which fragments separated in Q1 can interact with a heavy gas, such as xenon, and further fragment for separation in Q3. Used primarily for studying fragmentation mechanisms.

Turbomechanical pump   High vacuum pump that uses a series of vanes mounted on a shaft. They rapidly rotate between stator plate entraining air molecules and dragging them out of the evacuated volume. The "turbo pump" operates like a jet engine to evacuate the mass spectrometer to the high vacuum needed for operation (10^{-6} or 10^{-7} Torr).

APPENDIX E

GC/MS SELECTED READING LIST

E.1 JOURNALS

1. *American Journal of Mass Spectrometry*
2. *American Laboratory*
3. *Analyst*
4. *Analytical Chemistry*
5. *Environmental Science and Technology*
6. *LC/GC Magazine* (United States and Europe)
7. *Rapid Communications in Mass Spectrometry*

E.2 BOOKS

1. Hans-Joachim Hubschmann, *Handbook of GC/MS: Fundamentals and Applications*, Wiley-VCH, Weinheim, Germany, 2001.
2. Petra Gerhards, et al. *GC/MS in Clinical Chemistry*, Wiley-VCH, Verlag, Germany, 1999.
3. Jehuda Yinon, *Advances in Forensic Applications of Mass Spectrometry*, CRC Press, Boca Raton, FL, 2004.

GC/MS: A Practical User's Guide, Second Edition. By Marvin C. McMaster

4. Eric Stauffer, Julia Dolan, and Reta Newman, *Fire Debris Analysis,* Academic Press, London, 2007.

5. Reta Newman, M. W. Gilbert, and K. Lothridge, *GC–MS Guide to Ignitable Liquids*, CRC Press, Boca Raton, FL, 1997.

6. Pascal Kintz, *Analytical and Practical Aspects of Drug Testing in Hair*, CRC Press, Boca Raton, FL, 2006.

7. F. W. McLafferty and F. Turecek, *Interpretation of Mass Spectra*, 4th ed., University Science Books, Mill Valley, CA, 1993.

8. J. Throck-Watson, *Introduction to Mass Spectrometry*, 2nd ed., Raven Press, New York, 1985.

9. W. McFadden, *Techniques of Combined Gas Chromatography/Mass Spectrometry: Applications in Organic Analysis*, Wiley-Interscience, New York, 1973.

10. R. R. Freeman, *High Resolution Gas Chromatography*, 2nd ed., Hewlett-Packard, Palo Alto, CA, 1981.

11. D. Ambrose, *Gas Chromatography*, Van Nostrand Reinhold, London, 1971.

12. M. C. McMaster, *HPLC: A Practical User's Guide,* 2nd ed., Wiley, Hoboken, NJ, 2007.

13. M. C. McMaster, *LC/MS: A Practical User's Guide*, Wiley, Hoboken, NJ, 2005.

INDEX

3-dimensional data array, 9, 12
4 segment tuning, 123
5 level quantitation, 97
502 cal gas fragment, 68, 91,
70eV electrons, 42, 48

Abrasive source cleaning, 89
Accelerants analaysis, 111
Accurate molecular weights,
    127, 130
Acid extractables, 102
A/D board, 75
Adducts, 43
Air leaks, 48–49
Alpha effect, 34
Aluminum oxide paper, 87
Amu offset lens, 45
Analyzer rods, 44
Ansi-CDF format, 82
Annual sales, 15–16
Archived data files, 4, 82
Arson investigation, 111
Astrochemistry, 111
Atmospheric gases, 111

Atmospheric pressure interface (API), 18,
    138–139
Autotune, 50, 69, 156
Autosampler, 27–28, 152
Axial modulation, 123

Background, 165–166
Bar code vial I.D., 27
Base neutral compounds, 100
Base peak, 76, 167
Battery operation, 145
Benzidine calibration standard, 102
BFB target tuning, 69, 72, 95, 97
Blanks, 53, 74, 78, 105
Bomb sniffers, 111, 148
Breath analyzer, 146
Buffer, 139
Burn, 88–89
Burst fragment pattern, 132
Butterfly valve, 16, 39

Ceramic collar, 42, 91
Cal gas, 68
Calibrate, 50, 156
